T0010335

Praise for *Dirt to Soil*

"Gabe Brown's *Dirt to Soil* could not be more timely, as farmers are beginning to see an increase in costs of the fertilizers and many other inputs
they rely on. Gabe provides us with his complete story of how he transi-
tioned to a largely self-renewing and self-regulating (regenerative) farming
system. Even though I have been an organic farmer for 40 years, I was
amazed at how much I learned from reading *Dirt to Soil*. I highly recommend it to *all* farmers and food entrepreneurs, especially anyone interested
in anticipating future changes and preparing for them in advance."
　　—**Frederick Kirschenmann**, distinguished fellow, Leopold Center for
　　　　Sustainable Agriculture; author of *Cultivating an Ecological Conscience*

"Restoring the productivity of agricultural land is one of the most urgent
imperatives of our time. In this landmark book, Gabe Brown explains, step
by step, how farmers and ranchers can transform lifeless dirt to healthy
topsoil, offering a profound yet elegantly simple blueprint for reversing
land degradation across the globe."
　　　　　　　　　　　　　—**Dr. Christine Jones**, soil ecologist;
　　　　　　　　　　　　　　　founder of amazingcarbon.com

"Civilization was made possible by agriculture developed over the centuries
by ordinary people domesticating plants and animals using the emerging
biological sciences. Today mainstream agriculture—dominated by monoculture cropping and confined animal feeding—is the most destructive
industry ever to evolve. Based on chemistry and marketing of technology,
current agricultural practices produce twenty times more dead, eroding
soil than food, year after year. In this dangerous time, Gabe Brown's book
comes as a breath of fresh air, showing by example what any farmer who
cares enough about the future can do by following sound ecological principles and using common sense and imagination."
　　　　　　　　　　　—**Allan Savory**, president, Savory Institute

"This book is a must read for anybody who thinks that the 'Green
Revolution' has been a success. Gabe Brown provides a heartfelt personal
account of his journey and awakening to a new perspective on the importance of soil biology and the urgent need for a return to regenerative

integrated organic farming methods, not just to feed the world but also to save the ecosystem from imminent disaster."

—**Stephanie Seneff**, senior research scientist,
MIT Computer Science and Artificial Intelligence Laboratory

"*Dirt to Soil* skillfully describes the learning process and rich rewards of perseverance in the conversion from yield-driven farm practices that degrade soils to the regeneration processes that provide pride, productivity, nutrition, health, and sustainability to the basic infrastructure of society—agriculture. The core values of stewardship Gabe Brown describes for managing the agricultural ecology are reinforced by science that links diverse components so they function together to benefit everyone and everything in the dynamic rejuvenation of soils. The principles are exemplified through firsthand experiences that not only explain what, why, and how things need to change, but also provide the motivation to start doing them. The book provides hope that nutrition and health can be guiding principles in food production for successive generations to displace the 'Band-Aid' interventions adopted by the past two generations that have resulted in serious, unintended negative consequences."

—**Don M. Huber**, emeritus professor of
plant pathology, Purdue University

"Gabe Brown's dirt-to-soil story is an inspiring example of how land can flourish when a farmer tunes out the textbooks and chemical purveyors and starts listening to nature. Brown has become a folk hero in regenerative agriculture circles, and this book delivers his trademark candor and ability to cut through myth, jargon, and generations of bad advice to reveal essential dynamics of how farm ecosystems work. By keeping it real, this practical, spirited, and timely book can help spark an agricultural shift from conventional wisdom to common sense."

—**Judith D. Schwartz**, author of
Cows Save the Planet and *Water in Plain Sight*

"After hearing a presentation by Gabe Brown, many people picture Brown's Ranch in North Dakota as some sort of Xanadu where nothing goes wrong. This book provides a realistic perspective on Gabe's struggles in a challenging environment. Gabe and his family didn't let the challenges defeat them; they viewed them as opportunities to learn and innovate. This determination has made Gabe one of the leaders in the movement to regenerate soils. He has also helped to push, pull, and drag science toward finding solutions to solve our current farm and food crises. Farmers and ranchers like Gabe Brown and the others whose stories are told in *Dirt to Soil* are addressing the critical need to develop resilient systems that provide abundant, nutrient-dense food on regenerated soils that efficiently cycle nutrients and water through diverse biological communities."

—**Kris Nichols**, PhD, soil microbiologist,
KRIS Systems Education and Consultation

"Gabe Brown's story is a journey of hope and freedom for all those who care about food, health, and the earth. His passion to heal the land and serve others has shaken the foundations of the industrial agriculture model. The industrial agriculture complex is an insatiable furnace of consumption that devours soil, leaves farm families destitute, and impoverishes rural communities—ultimately destroying nations. Who would have thought that a North Dakota rancher would lead the regenerative agriculture revolution, a campaign that shows us a new way of growing nutritious food—food that is medicine and that nurtures and honors all of creation. Because of Gabe Brown, I have hope for the future of agriculture. *Dirt to Soil* is a must read!"

—**Ray Archuleta**, "The Soil Guy," retired
USDA/NRCS soil health specialist

"If you have interest in your health or saving the earth or eating food that tastes the way food should, you have heard a lot recently about regenerative farming. There are dozens (or hundreds) of self-proclaimed experts on the subject. Here is what I know: Gabe Brown is the Real Deal. He has done more than *anyone* to bridge the gap between research scientists and practicing farmers. His understanding of how to put the science of soil regeneration into practice is unsurpassed. *Dirt to Soil* should be required reading for every industrial farmer on the face of the earth."

—**Will Harris**, White Oak Pastures, Bluffton, Georgia

"*Dirt to Soil* is the perfect title for this new book from Gabe Brown. It is an apt metaphor to describe the Brown family's escape, through intelligence and determination, from their entrapment in an agricultural model that has failed economically, ecologically, and as a source of quality of life for the family. Their accomplishment stems from their realization that long-term success is possible only when all parts of the soil-plant-animal-wealth-human complex we call a farm/ranch are nurtured simultaneously. The Browns understand that agriculture must be about promoting life. It must be regenerative."

—**Walt Davis**, author of *How to Not Go Broke Ranching*

"I can no longer drive by a farm without wanting to get out of the car and start planting cover crops. *Dirt to Soil* is an entertaining, illuminating read that will change the way people think about agriculture."

—**Mark Schatzker**, author of *The Dorito Effect*

"There is growing awareness that industrialized agricultural methods are doing considerable damage to our soil, farms, and planet. The way we farm needs to change, and *Dirt to Soil* is about to transform the way agriculture is practiced around the world. This book is filled with excellent advice on how to farm by following Mother Nature's guidance, seasoned with Gabe's great sense of humor and humility."

—**Colin Seis**, agricultural management consultant; owner of Winona, New South Wales, Australia

"Reading *Dirt to Soil* is like having a personal conversation with Gabe Brown about the changes he witnessed as he put regenerative practices into place on Brown's Ranch. Most important is his clear message about capturing the value of his system by marketing directly to the consumer. In Gabe's words, he 'prefers to sign the back of the check, not the front.'"

—**Dwayne Beck**, PhD, manager, Dakota Lakes Research Farm; professor of plant science, South Dakota State University

"Civilization after civilization plowed itself out of prosperity by degrading the soil. Gabe Brown's inspiring story shows why regenerative farming to rebuild healthy, fertile soil is not just an academic theory—it's already been done on farms like his."

—**David R. Montgomery**, author of *Dirt* and *Growing a Revolution*

"*Dirt to Soil* is an inspiration! Gabe Brown offers a proven model that will help other farmers improve their soils and our planet. More and more growers are starting to realize that they must be *soil* farmers first and foremost. Gabe helps us imagine what the health of our planet would look like and how the profitability of farms, both large and small, would improve if all of us shift the way we practice agriculture. His message to mimic nature rather than trying to fight against her resonates with everyone who cares about the quality of our food and our future."

—**Todd Colehour**, founder, Williams and Graham and Tribe Market

Dirt
to
Soil

Dirt
to
Soil

One Family's Journey into
Regenerative Agriculture

Gabe Brown

Chelsea Green Publishing
White River Junction, Vermont
London, UK

Copyright © 2018 by Gabe Brown.
All rights reserved.

Unless otherwise noted, all photographs by Gabe Brown.

No part of this book may be transmitted or reproduced in any form by any means without
permission in writing from the publisher.

Project Manager: Sarah Kovach
Editor: Fern Marshall Bradley
Copy Editor: Deborah Heimann
Proofreader: Angela Boyle
Indexer: Nancy Crompton
Designer: Abrah Griggs

Printed in Canada.
First printing September, 2018.
14 13 12 11 10   23 24 25 26 27

Our Commitment to Green Publishing
Chelsea Green sees publishing as a tool for cultural change and ecological stewardship. We
strive to align our book manufacturing practices with our editorial mission and to reduce the
impact of our business enterprise in the environment. We print our books using vegetable-
based inks whenever possible. This book may cost slightly more because it was printed on paper
that contains recycled fiber, and we hope you'll agree that it's worth it. *Dirt to Soil* was printed
on paper supplied by Marquis that is made of recycled materials and other controlled sources.

Library of Congress Cataloging-in-Publication Data
Names: Brown, Gabe, author.
Title: Dirt to soil : one family's journey into regenerative agriculture / Gabe Brown.
Description: White River Junction, Vermont : Chelsea Green Publishing, [2018]
  | Includes index.
Identifiers: LCCN 2018023416| ISBN 9781603587631 (pbk.) | ISBN 9781603587648 (ebook)
Subjects: LCSH: Organic farming—North Dakota. | Integrated agricultural
  systems—North Dakota.
Classification: LCC S605.5 .B765 2018 | DDC 631.5/8409784—dc23
LC record available at https://lccn.loc.gov/2018023416

Chelsea Green Publishing
White River Junction, Vermont, USA
London, UK
www.chelseagreen.com

To the shy farm girl who wanted to marry a city boy to take her away.
Instead, that city boy brought her back to the farm
where she dedicated her life to being my number-one supporter
on this journey of regenerating dirt to soil.

A special thank you to our children, Kelly and Paul.
Your unwavering love and support is all a father could want.

Contents

Preface

I met Gabe Brown for the first time when I invited him to speak at the 2012 Quivira Coalition conference, which was titled "How to Feed Nine Billion People from the Ground Up." The theme for the conference sprang to my mind the previous year while I was visiting Colin Seis, a sheep farmer in New South Wales, Australia. We had fallen into an intense discussion about pasture cropping, an innovative type of regenerative agriculture Colin and neighbor Darryl Cluff had pioneered that involves growing annual crops and perennial pastures together. As we talked, I realized Colin and Darryl's work offered an intriguing solution to the rising challenge of sustainably supporting the estimated nine billion people that will be living on planet Earth in 2050.

Pasture cropping is one answer, and so are other practices that are part of regenerative agriculture, a biological system for growing food and restoring degraded land. Its goal is to continually advance the health of the soil with practices that promote microbial activity, increase carbon cycling, and improve plant and animal health, nutrition, and productivity—all of which can support feeding a lot of people. Practices include no-till farming, diverse cover crops, multiple crop rotations, on-farm fertility, minimized use of herbicides, and avoidance of all use of pesticides, insecticides, and synthetic fertilizers. All of that is also integrated with managed livestock grazing. As Colin Seis has demonstrated on his farm, regenerative agriculture can be profitable, too.

Colin and I grew excited over the prospects of the conference. When I asked him for suggestions about other leaders in regenerative agriculture who would be good speakers, the first person he mentioned was Gabe Brown.

As our audience learned at the Quivira conference, Gabe and his wife, Shelly, purchased a farm near Bismarck, North Dakota, from Shelly's parents in the early 1990s and began growing grains and raising beef cattle the conventional way with heavy tillage and plenty of herbicides, insecticides, and synthetic fertilizer. Three years later, they stretched the standard farming model a bit by switching to no-till practices in order to conserve soil moisture and reduce fuel costs. However, four successive years of weather-related crop failures created a desperate financial situation that set the Browns on an unexpected and revolutionary journey from industrial agriculture to biological, regenerative farming.

Their 5,000-acre ranch, Gabe told the conference-goers, now profitably produces a wide variety of cash crops, such as corn and wheat, and cover crops. Gabe grows cover crops throughout the growing season to address resource concerns such as protecting the soil. Brown's Ranch also produces grass-finished beef and lamb, along with pastured laying hens, broilers, pigs, honey, vegetables, and fruit, all marketed directly to consumers. What many conventional farmers and ranchers view as major challenges—such as soil compaction, wind erosion, flooding, diseases, pests, weeds, high input costs, and low yields—Gabe sees as symptoms of a poorly functioning ecosystem. The Brown's Ranch model, developed over twenty years of experimentation and refinement, addresses these resource concerns in a variety of ways, but a critical piece is focused on regenerating the living biology in the soil.

Gabe's talk at the Quivira Coalition conference proved so inspiring and popular that I invited him back in 2014, along with his son, Paul, to teach a workshop.

One of the eye-opening topics they covered in that workshop was how to grow topsoil. Grow soil? According to conventional thinking, it takes a thousand years to grow one inch of topsoil. Using regenerative agricultural principles, however, Gabe said, they had grown several inches of new topsoil in only twenty years! Through a synergistic combination of soil microbes, mycorrhizal fungi, earthworms, organic material, plant roots, water, sunlight, and the "liquid carbon" plants create via photosynthesis, they rediscovered a natural process for transforming the compacted, depleted dirt of industrial farmland into rich, porous soil. The reason for this transformation is simple, Gabe and Paul told the packed room: Biological life is a force, and once it has been unleashed, it will continue to grow and generate new life.

Not surprisingly, Gabe has become a popular, well-traveled public speaker and advocate for regenerative agriculture. In the winter of 2016–2017 alone, he gave more than one hundred presentations, speaking to over twenty-three thousand people, not to mention the 250,000 views his presentations have enjoyed online. Hundreds of people visit Brown's Ranch each summer, and many more visit the ranch website. Gabe has been featured in a number of documentaries on food and soil health in recent years—all evidence, he believes, of a groundswell of interest in regenerative agriculture taking place among consumers, ranchers, farmers, and even conventional producers who want to make a change.

What was missing, however, was a book. The editors at Chelsea Green had been encouraging Gabe to write about his experiences, but he found that finding time to do so proved elusive. I became involved in the project as a result of a chance conversation with Fern Marshall Bradley, a senior editor at Chelsea Green. We agreed that a book by Gabe would be valuable to the cause of regenerative agriculture. I asked if there was some way I could help bring the book into existence. Gabe was open to a collaboration, and a few months later we set to work, with my job mostly being a "word wrangler." I am honored to be involved in this project, and I am just as inspired today by the Browns' work as I was when I first met them.

In this era of hyper divisiveness, virtual realities, and baffling disdain for facts, Brown's Ranch demonstrates that we can be united by our common need for healthy soil. There is nothing virtual about growing food. You can't eat pixels. Your body needs nourishment, which means we need farms and ranches, which need soil. If we want to be healthy, then we need healthy food produced from healthy soil—not dirt—which we can accomplish only via biology, not chemistry. If we want to heal divisions, be resilient, and create opportunities for our children, then we need to start with soil and work our way up, one plant and one animal at a time.

It can be done, as the Browns show, if we set our minds to the task.

—COURTNEY WHITE

Introduction

The Best Teacher

Our lives depend on soil. This knowledge is so ingrained in me now that it's hard for me to believe how many soil-destroying practices I followed when I first started farming. I didn't know any better. In college I was taught all about the current industrial production model, which is a model based on reductionist science, not on how natural ecosystems function. The story of my farm is how I took a severely degraded, low-profit operation that had been managed using the industrial production model and regenerated it into a healthy, profitable one. The journey included many trials and constant experimentation, along with many failures and some successes. I've had many teachers, including other farmers and ranchers, researchers, ecologists, and my family. But the best teacher of all is nature herself.

In the everyday work of my farm, most of the decisions I make, in one way or another, are driven by the goal of continuing to grow and protect soil. I follow five principles that were developed by nature, over eons of time. They are the same anyplace in the world where the sun shines and plants grow. Gardeners, farmers, and ranchers around the world are using these principles to grow nutrient-rich, deep topsoil with healthy watersheds.

The five principles of soil health are:

1. **Limited disturbance.** Limit mechanical, chemical, and physical disturbance of soil. Tillage destroys soil structure.

It is constantly tearing apart the "house" that nature builds to protect the living organisms in the soil that create natural soil fertility. Soil structure includes aggregates and pore spaces (openings that allow water to infiltrate the soil). The result of tillage is soil erosion, the wasting of a precious natural resource. Synthetic fertilizers, herbicides, pesticides, and fungicides all have negative impacts on life in the soil as well.

2. **Armor.** Keep soil covered at all times. This is a critical step toward rebuilding soil health. Bare soil is an anomaly—nature always works to cover soil. Providing a natural "coat of armor" protects soil from wind and water erosion while providing food and habitat for macro- and microorganisms. It will also prevent moisture evaporation and germination of weed seeds.

3. **Diversity.** Strive for diversity of both plant and animal species. Where in nature does one find monocultures? Only where humans have put them! When I look out over a stretch of native prairie, one of the first things I notice is the incredible diversity. Grasses, forbs, legumes, and shrubs all live and thrive in harmony with each other. Think of what each of these species has to offer. Some have shallow roots, some deep, some fibrous, some tap. Some are high-carbon, some are low-carbon, some are legumes. Each of them plays a role in maintaining soil health. Diversity enhances ecosystem function.

4. **Living roots.** Maintain a living root in soil as long as possible throughout the year. Take a walk in the spring and you will see green plants poking their way through the last of the snow. Follow the same path in late fall or early winter and you will still see green, growing plants, which is a sign of living roots. Those living roots are feeding soil biology by providing its basic food source: carbon. This biology, in turn, fuels the nutrient cycle that feeds plants. Where I live in central North Dakota, we typically get our last spring frost around mid-May and our first fall frost around mid-September. I used to think those 120 days were my whole growing season. How wrong I

was. We now plant fall-seeded biennials that continue growing into early winter and break dormancy earlier in the spring, thus feeding soil organisms at a time when the cropland used to lie idle.

5. **Integrated animals.** Nature does not function without animals. It is that simple. Integrating livestock onto an operation provides many benefits. The major benefit is that the grazing of plants stimulates the plants to pump more carbon into the soil. This drives nutrient cycling by feeding biology. Of course, it also has a major, positive impact on climate change by cycling more carbon out of the atmosphere and putting it into the soil. And if you want a healthy, functioning ecosystem on your farm or ranch, you must provide a home and habitat for not only farm animals but also pollinators, predator insects, earthworms, and all of the microbiology that drive ecosystem function.

Throughout this book I return to these principles over and over again. I even devote a chapter to discussing their importance in depth (chapter 7). They are ingrained in everything I do on my ranch. It is my hope that, by the time you finish reading this book, you will not only know them by heart, but you will want to take advantage of them to regenerate your ecosystem, too. This is the journey of *Dirt to Soil*.

PART I
The Journey

One

Lessons Learned the Hard Way

How did a guy who grew up in the city, whose only contact with growing plants was mowing lawns during the summer, become so committed to soil health and land regeneration? It's a question I sometimes ask myself as I consider all that land regeneration has blessed me with.

I grew up in Bismarck, North Dakota, the third of four sons born to a father who had a lifelong career with the local rural electric cooperative, and a mother whose main job was to keep four boys out of trouble. My childhood was relatively uneventful, involving a great deal of baseball, bowling, and homework but not much exposure to agriculture except for a few brief trips to an uncle's farm. That all changed in the ninth grade when I took a class in vocational agriculture, inspired by an older brother, Jay. Soon after, I joined the Future Farmers of America and quickly became infatuated with all things related to farming and ranching. I wanted to learn everything I could, which in those days meant the where, why, and how of fertilizers, pesticides, insecticides, fungicides, artificial insemination, feedlots, balancing rations, diesel engines, and anything else related to industrial agricultural production.

During high school, I spent after-school hours picking rocks out of cropland fields for a local farmer, which is not an uncommon thing to do in North Dakota. This was the first time I had actually worked on a farm, and despite the rocks, I loved it. Little did I know that the farmer would soon become my father-in-law! My sweetheart, Shelly, and I were married in 1981.

My in-laws, Bill and Jeanne, were tremendously hardworking people who had started out in 1956 with little more than a dream, and through years of dedicated labor, eventually paid off a 1,760-acre farm while raising three daughters. In 1983, after I studied agricultural economics and animal science in college, Shelly and I moved into a trailer house on Bill and Jeanne's farm. They had asked us if we would be interested in eventually taking over the operation. Of course, we were eager to do so! Oops, I should rephrase that because Shelly insisted she married a city boy in order to get off the farm. And there I was leading her back to it! She must have loved me, though, because she didn't say no. Her parents farmed the land until 1991. With no son to take over, they had to settle for a son-in-law who had grown up in town and had little farming experience.

Bill and Jeanne farmed conventionally, including heavy tillage. In fact, I often tell people that my father-in-law practiced "recreational tillage." He loved to sit on a tractor and pull a heavy disc through the field. Every year they would rest half of the cropland for the summer, a practice called *fallowing*, tilling it repeatedly to keep weeds from growing. They fallowed their land because they thought it was a way to store moisture for the crop-growing year. On the other half of the land they would grow a cash crop. They grew small grains, mostly spring wheat, oats, and barley, and they fertilized annually, though not at heavy rates. They also used herbicides annually in the fields to kill weeds. They owned a sixty-five-head cowherd along with about twenty yearling heifers. These cattle were divided into three groups and then grazed in three native grass pastures on the farm for the entire growing season every year, year after year, without any variation. In the fall, the cattle grazed on the post-harvest crop residue and were then fed hay in a lot for five to six months during the winter. The calves were weaned in October and also fed for some time before they were sold. Their animals were subjected to the standard combination of pour-on insecticides and multiple vaccinations annually.

Bill and Jeanne sold all of their cattle in 1978. They rented out their pastures until Shelly and I moved to the farm and purchased our first group of registered Gelbvieh cattle, a breed that originated in Europe and was first imported to the United States in the 1970s. Gelbvieh are known for their milk, muscle, and mothering ability, and I saw them as a perfect fit for our ranch.

Becoming a Farmer

As I worked alongside Bill in my first years as a farmer, I learned about the conventional production model of agriculture. Even in the beginning I had questions about its logic. For example, in the spring Bill and I would till the soil, and I remember him telling me that we were "working the soil in order to dry it out." That didn't make sense to me, because in July we were always praying for rain! I distinctly remember him telling me "the more you work the soil, the better it is!" *Why?* I would ask myself. I tended to question his judgment from time to time, which didn't sit too well with his stubborn German ancestry. It was a good experience for me, however, especially as I had begun to make plans for things I would change after Shelly and I purchased the farm. Shelly has since admitted that it was a stressful time for her because she would have to listen to me complain about her parents in one ear and then listen to her parents complain about me in the other.

Our livestock management was conventional, too. During the growing season we looked at only three things: the cows, the grass, and the water. But after I met a rancher named Ken Miller, who was doing things quite differently, I began to question our grazing methods as well. Ken was, and still is, a mentor to me. He and his wife, Bonnie, ranch in a pretty tough environment near Fort Rice, North Dakota. The soils there are composed of a high percentage of bentonite clay and usually do not grow enough grass to keep a prairie dog fed—except on their ranch. By tenacious observation and careful management, Ken and Bonnie have healed their land to the point where it is extremely productive and profitable. Ken taught me things that my college professors never even mentioned. I am forever grateful to him.

To Bill and Jeanne's credit, they showed patience and allowed me to cross-fence a couple of the pastures so I could experiment with different grazing strategies. That was my first attempt at land regeneration—I just didn't know it at the time.

After I had worked alongside Bill and Jeanne for eight years, they made an unexpected decision to sell a third of the farm to each of their three daughters rather than sell the entire operation to us. That outcome was not what Shelly and I had worked toward. We had planned, and expected, to be able to purchase the entire operation. Twenty years later, this lesson weighed heavily in our decision of how we would transition the ranch to our own children, a decision I explain in detail in chapter 5.

We purchased the home place, comprising 629 acres, from Bill and Jeanne in 1991. We were fortunate to have the USDA's Natural Resources Conservation Service (NRCS) come out and do baseline tests on our soils. Two results were particularly important for our story. The first test showed that the percentage of organic matter in soils on our cropland ranged from 1.7 percent to 1.9 percent. I have since learned that soil scientists estimate that soil organic matter levels where I live were once in the 7 percent to 8 percent range. Approximately 75 percent of the organic matter that was once in my soils was lost over time due to tillage and improper management. When organic matter is depleted, the nutrient cycle in the soil is adversely affected. (This concept ties in with those all-important principles of soil health mentioned in the introduction and discussed in depth later, in chapter 7.) Many farmers turn to inputs of synthetic fertilizers to provide plants the nutrients they need. By the way, soils anywhere in the nation are typically composed of 50 percent minerals (sand, silt, and clay), 25 percent water, 15 percent air, and less than 10 percent organic matter (much less today).

The second test performed by NRCS on our place involved the rate at which rainfall could infiltrate our soils rather than ponding on the surface and evaporating or exiting the ranch as sheet flow. They determined the rate of water infiltration was a half inch per hour, which is typical for many operations in the area. The trouble was we needed every drop we could get. On average, our ranch received only sixteen inches of total precipitation per year, of which approximately eleven inches was rain and the remainder came from the seventy-plus inches of snow we normally got each winter. Worse, a large part of our rainfall came from thunderstorms, which could dump an inch or two of rain in a short time. A low infiltration rate meant most of that water ran off the land and thus was not available to plants. This presented a serious challenge in normal years, but it was especially difficult in periods of drought.

Looking back, I wish I had the foresight to archive some of those soils from 1991. It would be interesting to analyze them with today's technology and see just how degraded and devoid of life they were.

For the first few years after we purchased the home place, I continued to farm conventionally using tillage, fertilizers, and herbicides to grow small grains, as my in-laws had been doing. I simply did not know any other way; it was what I had learned in college and from Bill. Because I enjoyed livestock, I decided to increase my cattle numbers. I wanted some early season pasture, so I decided to seed 200 acres of cropland back to perennial grasses. Bill thought I was nuts. Why would anyone seed "nice" cropland back to grass? That was just not done! After a discussion with the local NRCS staff, I decided to use a seed mixture of smooth bromegrass, intermediate wheatgrass, and pubescent wheatgrass. A very good stand of perennial plants quickly established themselves on the former annual cropland, but it was not very productive. This eventually taught me an important lesson about how soil functions, as I explain in chapter 3. I tell people that I had to learn every lesson the hard way, and that lesson was one of the hardest!

Going No-Till

In 1994, a good friend of mine from the northern part of the state who was practicing no-till farming recommended that I switch to no-till practices in order to save time and moisture. His advice made sense. One of the benefits of not growing up on a farm was that I had an open mind. I did not have any preconceived notions. He added some sage advice along with his suggestion: "If you do go no-till, sell all of your tillage equipment so you are not tempted to go back." As a beginning farmer, I couldn't afford to simply up and buy a no-till drill, so I did what he advised. I sold all of my tillage equipment and, with the money I made, purchased a fifteen-foot John Deere 750 no-till drill. I ended up using that drill for twelve years before upgrading to a newer model.

While I was excited by the no-till process, Bill was extremely skeptical, especially after watching me seed into the previous year's residue. He was used to seeing finely tilled, bare soil and was tough to convince that no-till methods would work. My first year of no-till farming was fantastic. Not only did our crop yields go up, but I was also able to move down the path of reducing synthetic fertilizer use by adding nitrogen-fixing field peas to the crop rotation. I had learned that there are approximately 32,000 tons of atmospheric nitrogen above every acre of land, so I thought it was foolish to spend as much as I was on nitrogen fertilizer. Not only did the peas do well, but our spring wheat averaged 55 bushels per acre and sold for $4.58 per bushel. That was a very good price and an excellent yield at that time. I was on top of the world! I planted a winter triticale/hairy vetch mix that fall because I was looking for a crop I could cut for hay for my cattle. It germinated well and looked great. I thought farming was easy!

Little did I know the lessons I was about to learn.

Before I tell those stories, though, I'll tell you more about the no-till method and how it relates to soil health, which is the foundation that most of the stories in this book circle back to. Simply put, no-till farming is a practice that employs a seed drill, an implement with a single disc that cuts a slice in the soil no greater than the width of a knife. If there is crop residue on the field, the discs can easily slice through it. The drill has multiple seed units mounted on top. Seed drops through the center of each unit and is deposited in the small slit made by the disc at a pre-determined depth. The soil displaced by the disc is then gently pushed back into place over the seed by a closing wheel. The net result of a no-till approach is a field seeded with a cash crop, such as wheat, with essentially no soil disturbance.

What are its advantages? Tillage destroys soil structure, the home for soil biology, reducing water infiltration. With no-till farming, in contrast, there is more moisture available to grow plants due to increased rainfall infiltration as a result of better soil aggregation, increased organic matter, and adequate residue on the surface, which acts as a shield against evaporation. The potential for wind and water erosion is also significantly reduced. And by seeding into existing crop residue, no-till practices encourage the conditions for microbial life in the soil, which increases nutrient cycling and reduces the need for synthetic fertilizers. The process also requires fewer tractor passes over a field, reducing labor, fuel, and maintenance costs.

What are its disadvantages? Primarily, there is a lack of weed suppression, although all tillage eventually results in more weeds, often resulting in the increased application of herbicides to control those weeds. However, the use of herbicides disqualifies a no-till farm from being certified as organic, which can have economic impacts. No-till farming can also slow the warming of soils in springtime, which is necessary for germination. This slowed warming can be overcome by having a crop rotation that has correct carbon:nitrogen ratios, which I will talk about later.

The origin of the no-till movement can be traced back to Edward Faulkner, a radical agronomist and farmer from Ohio who declared in his 1943 book *Plowman's Folly,* "the truth is that no one has ever advanced a scientific reason for plowing."

In the Upper Midwest, no-till farming was introduced by Dr. Dwayne Beck, director of the Dakota Lakes Research Farm, near Pierre, South Dakota. Dr. Beck grew up on a grain and livestock farm in the area, received a degree in chemistry, and worked for a while at a fertilizer dealer before earning a PhD with a focus on soil fertility at South Dakota State University. When he began the research farm, his goal was to slow erosion on farm fields, which by the late 1980s had become a big issue, particularly on irrigated ground where large amounts of productive topsoil were washing away. While studying low-till and no-till systems for their conservation value, he observed dramatic increases in soil biology. He also noted that less water was needed to produce a profitable crop in these systems while both fuel and fertilizer usage dropped. When crops yields matched and then surpassed county averages, he knew the no-till method was the way to go.

The Dakota Lakes Research Station is owned by farmers, which allowed Dr. Beck the freedom to do things differently, including the pursuit of a systems-thinking approach to agriculture. He soon became an advocate for cover crops, which are plants grown primarily to enhance the life and the function of soil. He maintained that growing cover crops is the best way to create on-farm fertility. He set out to encourage no-till practices among local farmers in an area where heavy tillage had been normal for decades, convincing them one by one to give no-till farming a try. To counter the complaint that plowing was required to eliminate weeds, Dr. Beck told farmers that his research and experience showed that competition from a healthy cover crop took care of weeds. As for weed-killing herbicides—especially glyphosate, which was pushed hard by chemical companies—he predicted weeds would eventually develop a resistance to them. That prediction has come true!

Dr. Beck was one of the first in our region to insist that we look to nature for inspiration, especially the native prairie. As he has often said, "Nature never tills." What nature does is to develop diversity. In a prairie ecosystem, there are dozens of different species of plants, mostly perennials, growing together in symbiosis. Nature is an opportunist and abhors a vacuum—bare soil—and, if left undisturbed, nature will quickly increase both the quantity and diversity of plants. In a no-till system, a farmer can control the diversity of plants on a particular field. What you plant depends on your goals. If you have livestock, the plants can be forage crops. If you want to fix nitrogen in the soil, then you can plant legumes as part of the mix. According to Dr. Beck, no-till methods were practiced by the Native Americans in the area long before the settlers arrived. He recommended the book *Buffalo Bird Woman's Garden*, which tells the story of a Hidatsa woman in North Dakota who lived during the nineteenth century. In the book, she describes the no-till process as an indigenous practice, including raising thirteen different types of corn. It is an excellent example of growing food as part of nature. Today's farming is more like mining. Farmers excavate nutrients from the soil, including carbon, and haul it away. That's not a sustainable system, obviously.

The goal for a sustainable system is soil health. As Dr. Beck has said, *soil health* is a term that has been around since the 1990s, though for a long time it was hard to define. Today, we have a much better sense of what constitutes soil health, including the proper cycling of water and nutrients, the quantity of sunlight harvested, the diversity of biological life in the soil, how much carbon is being stored, and how resistant the soil is to erosion. Basically, how much is the soil acting like that of a prairie? But, according to Dr. Beck, there are no silver bullets, including no-till methods. An integrated, holistic approach is required in order to mimic the complexity and fertility of a prairie ecosystem on the farm.

There aren't any magic numbers either: There is no single indicator or test that will give a farmer the one number he or she needs to know to determine whether a soil is healthy. Imagine driving a vehicle down a road in a snowstorm—test result numbers are the white stripes at the edges of the road telling you when you're about to drift into a ditch. The farmer's job is to stay in the middle of the road as best as possible, despite the weather. Dr. Beck took this analogy a step further: How do you even know if you are on the right road in the first place? Do you know where it is leading? Do you have a map? What are your goals?

The Disaster Years

In the spring of 1995, I thought I was on the right road. In the crop fields, I planted peas, barley, oats, and 1,200 acres of spring wheat, all of which had synthetic fertilizer applied and were sprayed with an application of herbicide. Summer treated us well. I had hayed some of the winter triticale and hairy vetch, which I had seeded the previous fall, and I combined some for seed. Then, on the day before I planned to start harvesting the spring wheat, a devastating hailstorm claimed the entire crop. It was a total loss. My in-laws had farmed our land for thirty-five years, and only twice had hail damaged their crops, neither storm causing significant damage. Based on that history, I had not taken out hail insurance. I just hadn't thought it was a necessary investment. Boy, was I wrong! We were devastated.

Fortunately, the 150 registered Gelbvieh beef cows we owned were unharmed and we knew their calves, of which some were bulls, would provide some income. But with an operating loan and a mortgage to pay along with a young family to support, things were not going to be as easy as I had thought just a year earlier.

That fall, I decided to increase the number of acres planted to the winter triticale/hairy vetch mix, though due to a lack of money I did not fertilize any of those acres. Through the sale of bulls, steer, and heifer calves, and any other money we could scrape together, we were able to make the interest payment to the bank, but not the principal payment. I remember a gnawing feeling in my gut. How was I going to get out of this debt?

In 1996, we added corn to the crop rotation. We also seeded more acres of field peas, which we did not fertilize. Our banker was willing to stick with us, but he required us to purchase federal crop insurance. We didn't take out hail insurance, however. What a mistake this turned out to be, as a late July hailstorm once again wiped out our cash crops. Our hearts sank. Things had become very serious.

That fall and winter were tough. Our daughter, Kelly, had been diagnosed several years earlier with a serious case of scoliosis, which required her to wear a body brace. The brace was form-molded and needed to be enlarged as she grew. At age twelve, she was growing fast and this meant she needed a new brace every six months at a cost of several thousand dollars each. Insurance did not cover the braces. Both Shelly and I had taken off-farm jobs to help pay the bills, but at that point we had to pay back operating notes without a crop to sell, as well as deal with the mounting medical expenses. It all added up to a very stressful time.

I learned to live on very little sleep. I held down a forty-hour-a-week job during the day and farmed at night. Many a time I caught myself nodding off while driving the tractor. My father-in-law often commented about how crooked my rows were!

The good news was that we were earning a reputation for selling high-quality Gelbvieh bulls. They added income to the ranch, but we were of the mindset that we needed maximum growth and milk in these cattle. We used any inputs we could afford to accelerate the animals' growth, including implanting the steers, along with insecticides, wormers, vaccines, and the list goes on and on. When analyzing the cattle, we used a system called expected progeny differences (EPDs), which was a fairly new concept at that time. It involved tracking genetic markers that indicated preferred qualities in a bull or cow. I did not foresee that by following this system, I was sending our cowherd down the wrong path; this was another lesson it would take time to learn.

Meanwhile, 1997 came along. In early April, our 205 Gelbvieh cows were almost finished calving when a devastating blizzard hit. For three days we were pounded with record cold temperatures and snowfall accompanied with fifty-plus-mile-per-hour winds. I checked on the few cows left to calve every two hours, but it was difficult to even see them through the snow, let alone help them if the need arose. The second evening I headed toward the barn, which is located only three hundred feet from the house, but I could not see it because the snow was so heavy. I had made that trip hundreds of times over the years, but I could tell something wasn't quite right. Just then my foot caught on something and I tripped. I realized that I had walked to the side of the barn and my foot caught on the top of a board windbreak. The snowdrifts were so high that it had drifted completely over a ten-foot-tall windbreak! I picked myself up and headed back to the house. Saving a calf was not worth losing my life. I waited until daybreak before checking on the cattle. I clearly remember thinking, "This is crazy."

On the fourth day, the storm subsided, and Shelly and I went to work moving massive amounts of snow. The board windbreak that I had tripped over two days earlier was under four additional feet of snow. The drifts had reached the top of the barn—I have pictures of calves walking on top of the snow and standing on the top of the barn!

The first thing we did was search for the cattle. Our hearts sank once we found them. They had crowded together and many of the young calves had been trampled. That day we dug fourteen dead calves out of snow, and we found more in the following weeks. It was heartbreaking and frightening. We badly needed the income from those calves to pay back the mountain of debt from the previous two years of crop failures. Instead, the debt grew.

Our banker let us know we couldn't borrow any amount of money above what the surviving beef calves would bring on the commercial market. He also wouldn't allow us to figure any income from the cash grain above what federal crop insurance would guarantee, which, with two zeros in our proven yields due to the past two years of disasters, was not much. We were going to have to "farm lean." I planted a sizeable amount of our cropland acres with alfalfa with the intention of selling high-quality hay to dairies in Minnesota. I also planted a number of acres with sorghum/sudangrass mixed with cowpeas. I was moving toward cover cropping, although I wouldn't have known to describe it that way back then.

The weather that spring turned hot and dry. The heat continued through the summer, and by fall, growth was so poor we were not able to harvest a single acre of our cash crops. We were fortunate that we were able to scrape together enough hay to feed the cowherd, so they provided some income. The off-farm jobs, although barely above minimum wage, helped also. But, once again, the operating loan debt grew. Make no mistake, we were in a deep hole from which I was not sure we would ever emerge.

For the first time, I questioned my career choice—and my wife questioned her choice of husband (although this was probably not the first time she questioned that). Looking back, we laugh because anyone with any common sense would have quit!

The fortunate thing was that the land was on a contract for deed with Shelly's parents, which meant that if the bank called in our loan, they would not be able to sell our land. All they would get was a 4440 John Deere tractor, a John Deere 3020 tractor with an old F-11 loader, a square baler, a few miscellaneous items, and that John Deere 750 grain drill. They must have thought that the money they would receive from the sale of the equipment wasn't worth the effort of the paperwork, and seeing as how we were able to scrape enough together to once again pay the interest, they stuck with us.

Shelly had an uncle and aunt, Dan and Alice, whose farm was located only five miles from ours. We helped them out when they needed it, because they didn't have any children to do so. As they approached retirement age, we talked to them about the possibility of selling us their land. In 1997, they agreed to sell us 280 acres on a contract for deed. They were kind enough to allow us to purchase it with only a small down payment, which was crucial because we, obviously, did not have much. The land we bought consisted of 160 acres of native prairie and 120 acres that were enrolled in the USDA's Conservation Reserve Program (CRP), which takes land out of production for a ten-year period. There was one year remaining on the ten-year contract. That land was hilly with soils that had been eroded due to years of tillage prior to CRP enrollment. We purchased it with the intention of letting the contract expire and converting it to pasture. There were no fences or water on the CRP acres. Obviously, adding that infrastructure would have to wait until such a time as we could afford it.

Seeing Things Differently

Even during the disaster years, I read as much as I could about soils. I found Thomas Jefferson's journals particularly useful, including his ideas on how to grow crops without much money, because he, too, ran out of money at one point. When I read that he was growing vetch and turnips, I thought, "I can do that!" I read Lewis and Clark's journals about what they saw as they came up the Missouri River during their exploration. I read their descriptions of the prairies and the plants that grew here, and I learned even more about how this land evolved with grazing animals. I began to conceptualize the native rangeland on the farm as prairie that needed to be grazed by our livestock in accordance with nature's principles.

I had been studying Allan Savory's ideas about land management, too, and decided to cross-fence some of our ranch so I could create smaller pastures in order to move the cowherd more frequently. Savory, a biologist who studied the grazing behavior of wild animals in his native country of Zimbabwe, observed that herds would be constantly moving due to the pressure from predators. They often would not return to a given area for a long time. This resting period allowed the plant resource plenty of time to recover, which matched what I had read about our North American prairie soils. These soils had been formed with the help of large herds of bison who grazed heavily in one place but then moved on, leaving the land to rest and recover for the remainder of the year. This idea made sense to me,

Native or Not?

I had the privilege of hosting Allan Savory on our ranch, and he explained to me that I shouldn't refer to our pastureland as "native" since landscapes are ever evolving. He also explained that we shouldn't refer to particular plant species as "native" because species are ever evolving, too. We should refer to the landscape simply as an ecosystem of plants.

especially on our ranch's native pastureland, which, as far as we knew, had never been tilled.

Wanting to know more, during the winter of 1997–1998 I scraped together enough money to attend a "Livestock for Profit" conference in Bismarck. One of the speakers was Don Campbell, a rancher from Canada who practiced holistic management (the official name of Allan Savory's approach). Don made a statement I'll never forget; one I think about every day. Don said, "If you want to make small changes, change the way you *do* things. If you want to make major changes, change the way you *see* things." It was like a light went on in my head. Up until that time, I had been making only small changes on the ranch while praying for big results. I realized that I had to change the way I was seeing things. I needed to be looking at our entire operation differently if I was going to dig us out of the hole we were in and stay in business.

When the spring of 1998 blessed us with some rain, I was relieved. Perhaps our luck was changing. I decided to seed more acres with alfalfa. Since I didn't have much money for synthetics, I decided to fertilize only the corn acres and hope nature took care of the oat, pea, barley, and spring wheat acres. I sprayed most of the crops with herbicides in early June, and they looked good. All this changed one late June day when a sudden thunderstorm erupted into the all-too-familiar roar of hail. When the storm subsided, I surveyed the damage and found that over 80 percent of our crops had, once again, succumbed to the "great white combine." Was this really happening to us? How could anyone be this unlucky? It was a sad, sad time. Shelly wanted to quit; she had had enough. I was too stubborn. I could not stand the thought of being considered a failure. All I had ever wanted to do was ranch! I found solace in working harder, even longer hours.

If there was a silver lining from this hailstorm, it was that it occurred early in the growing season, which left me the opportunity to seed a forage crop. I chose sorghum/sudangrass and cowpeas. This crop grew well, but I couldn't cut and bale it for hay because I didn't even have enough money to buy twine. Instead, I left it standing and allowed the cattle to graze it during the late fall and early winter. Although I didn't know it at the time, this was my first attempt at winter grazing.

Four years, four disasters. The ironic thing is that none of our neighbors had suffered losses all four years. One suffered some losses during three of those years and several had two years of losses, but we were the only ones who had been decimated all four years. Was God trying to tell us something? Maybe we were too young, or dumb, or scared—probably some of each—to realize the trouble we were in, but I can honestly say that I never gave much serious thought to quitting. I had a college education and could have found a different career, but there was nothing I wanted to do more than ranch. Besides, I was too stubborn to give up. I wasn't going to give my neighbors the satisfaction of seeing the city boy fail. Today, I tell people that those four years of crop failure were hell to go through, but they turned out to be the best thing that could have happened to us because they forced us to think outside the box, to not be afraid of failure, and to work with nature instead of against her.

They sent me on this journey of regenerative agriculture.

Two

Regenerating the Ecosystem

The first clue that we were regenerating our ecosystem was the earthworms. I used to joke that we could never go fishing because there weren't any earthworms on our farm. Sadly, it was true. But after the four years of crop failures, I suddenly saw earthworms in the soil. It was as if a light turned on, and I began to realize what had happened. For four years, I had not removed our crops from the land other than the alfalfa we grew for the dairies. I had left all that biomass sitting on the soil surface, protecting it and feeding carbon to the microbes in the soil. I had also greatly reduced the amount of herbicide and synthetic fertilizer I used on the crops—because I couldn't afford it. The results were easy to see. I knew the soil was improving because when I sank a shovel in the ground, in addition to the earthworms, I saw darker, richer soil with better structure. It was beginning to change color and take the appearance of chocolate cake! This was a sign that the organic matter levels were increasing. The soil held more water, too. Even in a drought year, we had produced enough feed for our livestock because the health of the soil was improving.

I really knew I was on the right track, however, when I looked out a window of our house one evening and saw a pheasant fly by. That had never happened before! (Today, we have pheasants all over the ranch.) Deer, coyotes, and hawks were all showing up on our land as well. Some of this was due to the fact that Shelly and I had diligently planted hundreds of trees each year. Besides providing protection for our livestock, those trees provided a home and protection for the wildlife, too. But it fit a pattern: Life was returning to our ranch!

Regeneration

All of this set me to thinking. First, I realized that I had come to accept the degraded condition of our ranch as normal. Instead of reversing the degraded conditions, I had been trying to hang on and not let things become worse. I was trying to *sustain* the operation in a poor state of health, not help it recover and improve. I know *sustainable* is a popular buzzword today. Everybody wants to be sustainable. But my question is: Why in the world would we want to sustain a degraded resource? We instead needed to work on *regenerating* our ecosystems. Symptoms of a degraded resource included poor infiltration, poor fertility, compaction, weeds, low yields, high input costs, salinity, plant diseases, invasive pests, erosion, declining profits, and the list went on. The cause of all these symptoms was the same: poor ecosystem function. Thanks to the crop failures, I changed the way I looked at our land. Unfortunately, the Good Lord had to slap me four times before I woke up!

Second, I realized that the land was regenerating itself naturally. By not tilling for five years, by adding diversity—including nitrogen-fixing legumes—to the cash crops, by growing a cover crop, by leaving the biomass on the surface of the soil after the crop failures, and by nearly eliminating chemical inputs, I had created the conditions in which the soil biology could thrive again. In particular, the mycorrhizal fungi in the soil had had a chance to repopulate. These organisms form a symbiotic relationship with the roots of most plants and are essential to a healthy soil. Mycorrhizal fungi secrete a glue-like substance called glomalin that helps bind soil particles together, and the more soil particles, the more pore spaces. These pore spaces are critical for water infiltration, and it is in and on the thin films of water in soil pore spaces that most soil microbes live.

No matter what you do to the soil, there will still be some small bit of life in it, even in the most chemically dependent or heavily tilled operations. If you give that life a chance to grow, it will respond. That's what I realized when I suddenly saw the earthworms. If you build it, they will come—or in our case, if you stop destroying it, they will come. Fostering life is the key to transforming *dirt* into *soil*.

Although the term *soil health* was rarely used in the 1990s, I was beginning to see the elements of the *five principles of soil health* taking shape as we emerged from the years of crop failures. (I discuss these five principles in detail in chapter 7.) NRCS again visited our ranch during this time and retested our soils, and they found that the organic matter content had *increased* during those difficult years.

Our four years of crop disasters had turned out to be a blessing. Not only were we forced to rethink how we farmed, we gave the land a break from destructive industrial practices. I had done what I had to do to preserve my farming operation, and fortunately those things also created the right conditions for the soil to regenerate on its own. I didn't realize it at the time, but I was collecting more sunlight and cycling more carbon as a result, which in turn fed the microbes. We were beginning to heal our land, and I was no longer afraid of trying new things. From that point forward whenever I visited another farm or ranch, I paid much more attention to what did or did not work, always with the goal of figuring out new ways to advance my own operation.

Another opportunity for learning arose unexpectedly in 1998 when I was asked if I would consider running for a position on the Burleigh County Soil Conservation District board of directors. I agreed and was elected. It turned out to be one of the best deci- sions I have ever made. Jay Fuhrer was the district conservationist at the time, and he and I soon became good friends, as we both had a passion for learning. Finally, I had someone to bounce ideas off of. We spent every chance we could challenging each other. It was great! I am forever grateful to Jay for pushing my comfort level. I would not have been able to go down this path without that push. I ended up serving on that board for fourteen years and enjoyed every minute of it.

Improving Our Pasture Management

I also learned some important lessons about pasture management around this time. The first lesson concerned those 200 acres of tame grass pasture that I had seeded to perennials back in 1993. I had divided the acreage into eleven separate paddocks using a single strand of high-tensile wire. A lane down the center allowed the cattle to trail back to a water point. That turned out to be a big mistake, because late in the summer the trail ended up being bare soil from the repeated cattle traffic. The cattle kicked up dust as they walked, causing some calves to get sick. To solve this problem, I decided to try installing a shallow pipeline to transfer water to all the paddocks, so the cattle wouldn't have to trek to a single water point. I had not seen this done before and didn't know right from wrong, so I just did it.

My son, Paul, and I rolled out some cheap one-inch polyethylene pipe, fused it together with a splicer I borrowed from a contractor, and then proceeded to trench it in using a small trencher, which I also borrowed from a friend. It was a slow process, but we managed to bury half a mile in an afternoon. I installed one riser underneath every other cross fence. This allowed me to set a seven-hundred-gallon rubber tire tank under those cross fences. Each tank provided water to two paddocks. The tire tanks were permanent, providing the cattle with access to water in every paddock. (See plate 14 on page 7.)

People often ask me why I use rubber tire tanks. I explain that when you ranch next to tens of thousands of hunters, everything becomes a target during deer season. Whereas fiberglass or steel tanks would not hold up to a rifle slug, the bullets do not penetrate through the steel belts of the rubber tires.

To drain the waterline in the winter, I installed a petcock in the line at the lowest point in the pasture. In the fall I shut the water off, open the petcock, and insert a riser at the water tank near the highest elevation in the pasture. This allows air into the pipe, which forces the water to drain out of the petcock. I have been using this method for eighteen years and have never had a line break due to freezing. And, by the way, the frost will often reach six feet deep during our cold northern winters.

Another lesson I learned from these pastures concerned what was going on below the soil surface. Even though that original seeding gave me a good plant stand, it was not productive. The plants were spindly with little leaf area, and the leaves that did develop were small and thin. Very few plants produced seed heads. All of this pointed to a dysfunctional nutrient cycle. You see, back when I had seeded that stand, I was seeding into fields that had been tilled for many years. Those fields had seen little crop diversity. Remember, my in-laws seeded only spring wheat, oats, and barley—all cool-season grasses. Along with synthetic fertilizer, this was a recipe for failure. Mycorrhizal fungi had been destroyed over time by the till-age and fertilizer. The result was minimal soil aggregation, which meant there wasn't a home for soil biology and the water infiltration rate was very poor. All of that added up to a very poor environment for plant growth. The seeds germinated, but the plants were essen-tially suffering from starvation.

I should have addressed these concerns *before* I had planted the perennials. I sought the advice of several experts as to how to rectify this situation. Their answer: Apply synthetic fertilizer. After four years of little income, that was not an option. I knew I was going to have to solve this using only plants and animals. But which plants? I could not find any credible research that could tell me which species of grasses and legumes to use in my environment, so I decided I would experiment. I settled on ten different legumes. They included two different grazing alfalfa varieties, cicer milkvetch, birdsfoot trefoil, white and ladino clovers, sainfoin, and others. I sprayed each paddock with glyphosate to set the bromegrass back (I no longer use glyphosate—now when I seed into existing perennials I set back the perennials with a severe overgrazing). Then, I simply interseeded my chosen species—one species into each paddock, leaving one unseeded as a control. Due to the good rain in 1998, especially that which came with the hail, the new stand establishment was vigorous.

I observed those seeded paddocks over time, and from them I learned which species perform well and persist in my environment. For me, alfalfa, cicer milkvetch, and ladino clover work best. That may not be the case on your farm or ranch. You will have to experiment for yourself. Today, I seed mixes of forbs such as chicory and plantain, along with a diverse mix of grasses and legumes, into perennial forage stands. I never seed a monoculture perennial pasture. (My cover-cropping methods are covered in detail in chapter 8.)

The Long Climb out of Debt

Over the next few years, we saw a return to "normal" crop production. The weather was favorable, and our yields grew thanks to the healthier soil we had created by accident as a result of the years of crop failures. Although margins were barely above our cost of production, at least we were making a profit. Cash became available as a result, so we began using more synthetic fertilizer, though not at the levels we had prior to 1995.

We continued to sell registered bulls, and our reputation grew, bringing a decent profit. Like most other producers in the area, I turned bulls out with the cows in May in order to calve the following February and March. I began keeping a few bull calves to sell to other ranchers in the area as seed stock. I selected the best-performing bull calves at weaning time in October and castrated the rest, which we fed until January and then sold at the local auction barn. The heifer calves were also fed in our lots from weaning until January. As with the bulls, we kept the top performers and sold the culls through the sale barn.

I listened to industry experts and ran the cattle through the squeeze chute on a regular basis throughout the year. In January, the cows were vaccinated and wormed. During calving season, we treated any sick calves for pneumonia and scours. Before we hauled cow/calf pairs to pasture in May, we again wormed the cows and applied insecticide ear tags. We gave calves a respiratory vaccine and wormer. We held back those cows that we decided to artificially inseminate; they were given multiple shots and CIDRs (progesterone inserts) to synchronize their estrus. Once we had weaned them, we gave the calves a seven-way vaccine along with wormer. We boosted the vaccines with a second shot two weeks later.

As the demand for our bulls grew, we increased the number of cows. With herd expansion came more registration papers and more breeding herds (six at times) as we tried to match dams and sires that we thought would cross well. All of this ultimately led to more work, not only from all of the sorting and hauling of cattle to pasture but also in the form of photographing the animals, developing a sale catalog, and marketing the bulls. Things kept expanding, and in 2000 we went "big league" and moved our annual sale to the local auction barn in town to accommodate the growing number of bulls that we were offering.

We were constantly chasing performance and touted ourselves as offering the highest performance Gelbvieh bulls around. Eventually, we started crossing the bulls with Black Angus and Red Angus in order to take advantage of hybrid vigor. It was a trendy concept, and we were the first in the game to sell Balancer (Gelbvieh/Angus or Gelbvieh/Red Angus cross) bulls in North Dakota, which made our sales even more popular. To keep our reputation high, we added more steps to the bull production process. We weighed the animals multiple times throughout the fall (to see which were excelling in the feedlot), clipped them in December, and cleaned them multiple times throughout January and the early part of February to make sure they were spotless on sale day. Add to this the free delivery we offered, and it was a lot of work!

I began to dread the holiday season because I knew the day after Christmas I would have to begin bull cleaning and trimming (giving them a haircut). I realized that although I enjoyed many things about the bull business, it was taking me away from something I enjoyed much more . . . *my family*! I knew I had to find a way to change the production model.

Learning More About Mycorrhizae

In 2002, we had a field of dryland corn yield over 200 bushels per acre, which was unheard of in Burleigh County, North Dakota. When I was combining that field, my father-in-law asked my wife to take a picture of him and his grandson standing in it. I knew then that I finally had his approval.

I was proud of that corn, too, but I knew I had farther to go on the journey into regenerative agriculture. In 2003, I was fortunate to meet Dr. Kris Nichols, a soil microbiologist at the USDA Great Plains Research Station in Mandan, just across the Missouri River from Bismarck. Dr. Nichols's research interest was soil biology, but she came at it originally from an ecological perspective not a farming one. Her job was to study the natural processes that occur in the soil and one of her main interests was mycorrhizal fungi. There are different kinds of mycorrhizal fungi, and arbuscular mycorrhizal (AM) fungi are the key players in nutrient transfer throughout the soil. They are the streets and avenues of the soil. Fungi are one of the most prolific microbes on the planet, second only in quantity to bacteria. Composed of long, thin filaments, fungi are found in nearly every terrestrial ecosystem, although they are most commonly associated with woods and forests (think mushrooms).

In biology-based agricultural systems, the fungi grow in association with most types of crops, often densely. They act as an extension to the host plant root, extending far into the soil where they acquire needed nutrients for the plant in exchange for carbon compounds secreted by the plant from its roots. As a result, the presence of AM fungi vastly expands a root's access to minerals and other nutrients, which it can then absorb. The presence of AM fungi in association with roots can also stimulate a plant to generate antioxidants and phytonutrients internally. These compounds are critical to human health, too, as I explain in chapter 10. Although the chemical signaling between plants and fungi is less well understood by scientists than that between plants and bacteria, the mechanisms appear to be similar, and the net result—additional nutrient uptake by the plant—is clear.

Shortly after I met Dr. Nichols, I asked her to come out to our ranch to see what we were doing. After looking around, she said "Gabe, you've come a long way, but your soils will never be sustainable unless you remove your synthetic fertilizer inputs." She explained that synthetic fertilizer is detrimental to mycorrhizal fungi. By applying fertilizer, I was actually hurting my soil, not helping it. Synthetic fertilizer interrupts the relationship between microbes and plant roots because the fertilizer gives plants "free" nutrients, so they don't need to trade carbon for nutrients from microbes. When that happens, the plants keep a lot of that carbon for themselves, which means the microbes don't get enough food to grow and reproduce, and their populations suffer. Mycorrhizal fungi efficiently acquire minerals for plants in exchange for carbon, but if the fungi do not receive carbon from the plants, they can't acquire minerals, and then plants can't access those minerals. Also, synthetics supply only a limited type of nutrient, not the full range that plants need. (Don't forget, I was already getting nitrogen for free from the atmosphere thanks to the legumes in the cover crops.)

After hearing her advice, I thought: Wow! Are my soils healthy enough that I can farm without synthetic fertilizer? I decided to find out. Starting in 2004, I did split trials in several fields on our operation. In one half of the field, I put down synthetic fertilizer, albeit at a much a lower rate than recommended. On the other half of the field, I added no synthetic fertilizer. I decided to continue the trials for a four-year period in order to test the fields under varying weather conditions. The results turned out to be astonishing.

Cover Crop Cocktails

At the same time as I received Dr. Nichols's radical advice to drop synthetic fertilizer, I was learning more about the role of cover crops in regenerating soil. One important day was when I heard Dr. Ademir Calegari, a Brazilian agronomist who has traveled the world extensively teaching others how to advance soil health with the use of cover crops, speak at a No-Till on the Plains conference in Salina, Kansas. Two things in particular have stayed with me from Dr. Calegari's presentation. First, he said that whether your farm or ranch receives two inches of precipitation per year or two hundred inches, you can grow a cover crop. That's when I knew I could grow more covers. Producers tell me all the time they don't get enough precipitation to plant cover crops. It doesn't matter where they farm, whether it's someplace dry like North Dakota or someplace wet, people like to think their conditions are *too* dry or *too* wet. In reality, they're just making an excuse. That was Dr. Calegari's point—he'd been all over the world, and he'd seen cover crops grown in every sort of environment.

Second, and this point really struck me, he said cover crops are meant to be *seeded in multispecies combinations*. I remember immediately thinking, "Gabe, you idiot! That is how prairie ecosystems function!" Dr. Calegari explained that synergies compound when approximately seven or eight species are grown together in a "cocktail." Until then, I had only been using two- and three-way mixes on our operation, and most people thought I was crazy doing that. But in terms of ecosystem function, what Dr. Calegari said made a great deal of sense.

My good friend Jay Fuhrer, Burleigh County district conservationist at the time, attended that conference with me. When
we returned home, Jay proposed to the Burleigh County Soil
Conservation District board of supervisors that we run a demonstration test based on Dr. Calegari's advice. Little did we know what
an impact this test would have! The winter of 2005–2006 was very
dry, so there was virtually no soil moisture available in the spring. In
late May, soil district staff seeded 1-acre plots consisting of monoculture strips of eight different cover crop species. Four of these
species—cowpea, soybean, turnip, and oilseed radish—plus millet
and sunflower were also sown as a six-way cocktail blend (this is
called *polyculture*). Approximately one inch of rain fell on those
fields between seeding and late July. The results were astounding.
Production was *three times greater* on the plot with the diverse polyculture mix! A comparable area of each plot was clipped, air-dried,
and weighed to validate the results, even though they were obvious
just by observation.

Dr. Nichols explained the results this way, "Not only were the
[mycorrhizal] fungi providing for the needs of one plant, but the
fungal hyphae pipeline connects to multiple plants, thus supplying
both the nutritional and energy needs of both microorganisms and
plants." This synergy is how nature functions. Jay's demonstration
proved that monocultures are a detriment to soil health.

These tests made a huge impression on me, and I immediately
began planting eight-way, ten-way, and twelve-way cover crop blends.
Today, I rarely seed a cover crop blend with fewer than seven different
species, and most of the time I use more. I've seen the results. We've
greatly increased organic matter, improved soil health, and gotten
much greater production. The results of these side-by-side tests also
made a big impression on Jay. He told me the results didn't make
sense in terms of classical agronomic thinking, but once you started
thinking in terms of the diversity of nature, they made a lot of sense.

I could not stop thinking about how well the cover crop cocktail
performed in a drought. I knew the impact that cover crop blend

would have on soil health, but I was also about to find out how that impact could be magnified by yet another change in my methods.

The Power of Stock Density

Due to the successes were we having on Brown's Ranch, I'd begun receiving invitations to speak at agricultural conferences to share my story. In early 2006, I spoke at a forage conference in Manitoba, Canada, and after I was done, a short, bald guy with a long handle-bar mustache stormed up to me and said, "I need to show you what I'm doing! You need to come up to my room!" I was not about to go up to this guy's room, and I tried to cut off the conversation, but he was mighty persistent. He forged ahead with explaining how he was using *livestock* to change soils, and that caught my attention. As I listened, I became intrigued to the point where I relented, and I accompanied him to find his laptop. Once we got to his hotel room and started up his computer, he showed me photos of his ranch and the high-density livestock-grazing techniques that he used. We ended up talking so late into the night that his wife asked to borrow the key to *my* room so she could go and get some sleep. This has remained a joke between us ever since. And what I saw that night on that laptop provided a critical piece of the puzzle for me concerning regenerative agriculture, a piece that had a major impact on Brown's Ranch.

That Canadian rancher's name is Neil Dennis. He and his wife, Barbara, live on an 1,800-acre ranch near Wawota, Saskatchewan. No crops, just perennial pasture grazed by cattle. What I found unique was that Neil was using very high stock density grazing—he called it mob grazing—in order to regenerate his soils. Stock density refers to how many animals are placed on a given area of land. Low stock density is typical of many ranches, especially in dry country. The cattle are spread out over a large area and are usually not moved to new ground very often, if ever. Increasing the size of the herd or reducing the size of the paddock creates a shift to a higher stock density. If you do *both*, as Neil has done, the result is very high stock density. To avoid overgrazing the land, however, you must shorten the amount of time you allow the cattle to graze in any single paddock. In Neil's case, the amount of time can be as short as a few hours.

When my in-laws owned our farm back in the 1980s, their stock density was about 250 pounds (animal liveweight) per acre, with very little rotation. At the time I met Neil, I thought I was doing pretty well. My stock density was 50,000 pounds per acre, and I was moving the cattle to a new paddock once a day. But after listening to Neil, I realized that I wasn't even close to his stock density, which often reaches 800,000 pounds per acre! It sounds unbelievable, but photos don't lie, and his land looked great in those photos. No wonder we stayed up talking until three in the morning. I knew I had to see Neil and Barb's ranch for myself, so that spring, I drove to Saskatchewan. I have to admit, I was skeptical that moving cattle more than once a day would make much of a difference, but once Neil and I began digging holes in old cropland that Neil had let revert to pasture, I could immediately see how healthy his soil was.

The story of the transformation of Neil's ranch is similar to mine. His ranch had been in the family since the early 1900s, and it had been managed conventionally. By the 1980s, the ranch was facing financial difficulties. A friend of Neil's sent him a flyer about holistic management as an alternative way of managing his livestock and improving his financial situation, but Neil promptly threw the flyer into the trash can. His attitude was that it would never work on their ranch. At Barb's urging, however, Neil enrolled in a holistic grazing course. He argued with the instructor the entire time. He was still convinced it wouldn't work on his ranch, and his instructor said it would work, so he set out to prove his instructor wrong. Try as he might, Neil soon found out that his instructor was *right*. His land improved. Production increased and, along with it, so did profitability. That's when the light went on for Neil. He didn't have enough cattle on his land!

Holistic Management

Holistic is a Greek word that means all, whole, entire, total. In agriculture, holistic management is a systems-thinking approach to managing resources developed by Allan Savory.

The Savory Institute website states that holistic management is "a process of decision-making and planning that gives people the insights and management tools needed to understand nature: resulting in better, more informed decisions that balance key social, environmental, and financial considerations."

The premise of holistic management is that nature functions in wholes. It is a holistic community with a positive and mutualistic relationship between people, plants, animals, and the land. If you remove or change the behavior of any one of the keystone species (species that help define the characteristics of an ecosystem), it will have a wide-ranging negative impact on other areas of the environment.

Holistic management focuses on the four ecosystem processes and our potential impact on them. Those processes are: the *water cycle*, the *mineral cycle* (which includes the carbon cycle), *energy flow*, and *community dynamics* (the complex set of relationships of biology within the ecosystem).

Holistic management uses a set of testing questions to determine whether or not a proposed action takes you closer to or further from your holistic context:

Neil custom-grazes other people's cattle, and so he began to gradually raise the stock density on his land. To begin, he put one hundred cow/calf pairs on 20 acres. As he watched the grass thicken and the soil

1. **Cause and effect.** Does this action address the root cause of the problem?
2. **Weak link.** Would taking this action address the weak link in one of the following areas: social, biological, or financial?
3. **Marginal reaction.** Is there another action that could provide greater return for the time and money spent?
4. **Gross profit analysis.** Which enterprise(s) contribute(s) more to covering the overhead costs of the business?
5. **Energy/money source and use.** Is the energy or money to be used in the action derived from the most appropriate source to meet your goal?
6. **Sustainability.** If you take this action, will it bring you closer to or farther from the future resource base you desire?
7. **Society and culture.** How do you feel about this action now? Will it lead to the quality of life you desire? Will it adversely affect the lives of others?

On our ranch we use these testing questions for all major decisions. They make decision making much easier, and we are much more confident in our decisions and in the chances of success of the actions taken.

improve, he added more animals, raising the stocking rate gradually. The soil kept improving, so he kept adding cattle while reducing their grazing time in a paddock. Pretty soon he felt like he was conducting an experiment—how high could he go? He hit 500,000 pounds per acre, then 800,000. A few times, he even pushed his stock density to 1 million pounds! Everyone, especially government experts, told Neil he was crazy and it wouldn't work, which just made Neil work harder to prove them wrong. Neil likes to tell the story that, when he first started working with these methods, when he would walk into the coffee shop in town, all conversation would stop. It didn't bother him. As Neil likes to say, "You've got to stir the pot or it burns on the bottom."

You can imagine how Neil felt when, over time, the organic matter levels in his soil more than doubled. The water infiltration rate increased to sixteen inches per hour. And, by the way, the first person to notice the improvement was the man who graded the county roads. He stopped by one day and asked Neil why his ranch looked better than all the other operations along the road!

Here's a summary of Neil's advice about grazing:

- You need to mix things up. Don't graze the same paddock at the same time every year, and don't graze it with the same number of animals.
- Don't try to achieve high stock density all at once or in one place, the transition needs to be gradual.
- Start small. Try 5–10 acres and work up from there.
- Most importantly, learn from your failures. (I already knew that one!)

A key to Neil's success was the development of automated gate-release devices called Batt-Latches designed for use with electric fences. This was important for Neil because automation freed him up to spend more time observing the land and thinking up new things to try. Before the automation, repeatedly opening and closing gates to paddocks under high-density grazing with frequent movement of the herd involved a lot of traveling back and forth across the ranch. With the automation, his cows have learned to anticipate when the gate will open. He can set up a dozen automated gates to open in the proper sequence, and go on vacation. But Neil is having so much fun that he doesn't take vacation!

On the drive home from Neil's ranch that spring, I had already decided that we would use higher stock density on our pastures. Just after I crossed the US border, though, the realization struck me that livestock were the missing link on our cropland as well! Up to that point, I had limited our cattle to grazing on pastureland during the growing season, allowing some grazing on cover crops in the farm fields during late fall and winter. I wondered: What if instead of seeding a cash crop and only grazing it after harvest, I seeded a crop field to a multispecies cover crop and then grazed it *during* the growing season? Would I be closer to mimicking how my prairie soils were formed, with large herds of grazing ruminants (bison and elk) eating, trampling, and moving on? This was an unorthodox idea, especially since I was planning to use high stock densities as well. No one I knew was using anything like this approach to grazing, but after talking it over with Paul, I decided to try it. We had some cover crops growing already, and I decided to move cattle in, stocking them at 600,000–700,000 pounds of beef per acre and then rotating the herd quickly through the paddocks. It worked! I soon saw an improvement in the health of the soil. That's when I knew that integrating livestock into our whole operation was critical to creating a truly regenerative ranch.

One of the great things about holistic management is the flexibility it gives one to change as conditions change. Since forage growth and weather change constantly, planned grazing helps you keep pace as a manager. And with high stock density, you can use your livestock as a tool.

Another idea that Neil came up with is something he calls "deep land massage." He rolls out a bale of hay on a parcel of land that has too much bare soil, and as the cows eat it, they stomp some of the hay into the soil. This feeds the microbes, which in turn helps grow grass. Neil knows it works because he can see the difference

Bale Grazing

Although I do not use Neil's "deep massage" method to cover the soil, I do use a similar practice called bale grazing. Our first choice is *always* to have our livestock graze on standing forage. Anytime processed feed is fed, it will be more expensive than standing forage grazing. That being said, there are times when the ground is covered with so much snow or ice that it prevents the animals from grazing on standing forage. In those instances, we bale graze.

Bales can be set about fifty feet apart (there is no right or wrong here). Depending on the quality of the hay, most producers who bale graze give their animals access to about a weeks' worth of hay at a time. (They use a single-strand electric fence to make sure cattle only have access to that amount.) Once those bales are consumed, the cattle are given access to more bales. With this method, it is possible to start a tractor only once the entire winter to feed a cowherd, and that one time is to set the bales out on the pasture in the fall. The beauty of this system is that all the dung and urine are deposited directly on the pasture. There is no need to haul manure out of a corral. What a money saver! (See plate 15 on page 7.)

between those bare areas that have received the "massage" treatment and those that have not. It's like night and day.

Bare soil is one of the worst symptoms of a degraded ecosystem on any farm or ranch. It is the first step toward desertification in fragile areas that have limited rainfall throughout the year. And I don't mean just tilled soil, either. Bare soil in a pasture is also a sign of trouble, as well

Once we started bale grazing, we soon found that the benefits were numerous. In my northern environment it is common for livestock producers to feed hay for at least five months out of the year. Not only is it expensive either to put up your own hay or to purchase it, but it is very expensive and time consuming to feed it. I used to spend three to four hours each day for those five months in a loader tractor or a tractor pulling a feeder wagon, hauling feed to the cattle confined in corrals. This takes time and fuel and creates wear and tear on equipment. I spent time hauling hay to cattle in confinement and then more time hauling all of that manure out of the corrals and spreading it on fields. I had forgotten that my cattle had legs!

After we switched to bale grazing, in addition to the savings on labor, fuel, and equipment, we noticed a major impact on subsequent forage production in those paddocks where we bale grazed. The Burleigh County Soil Conservation staff came out and clipped, weighed, and tested the forage. They found that we tripled biomass production and the forage was significantly higher in protein as compared to our non-bale-grazed paddocks.

We use bale grazing extensively for both cattle and sheep. The dollars saved and benefit to the resource is immense.

as an opportunity for weeds to gain a foothold. Using cover crops is one method to reduce or eliminate bare spots, but animal impact is another. The prairies formed in the presence of huge herds of bison that would graze for only a few hours or days in any one place before moving on. Neil was recreating their effect, as I saw. How does it happen? All those hooves press a great deal of grass and other litter onto the soil, where it decomposes and becomes food for microbes. Add a large amount of urine and manure, and you have a lot of natural fertilizer.

Dr. Nichols explained it well: Animal grazing keeps plants in a vegetative state, which means the carbon produced by photosynthesis will stay below ground longer. Otherwise, the plant will recall carbon for use in seed production and further growth. Grazing can also stimulate exudate production (carbon secretions) in roots because physiologically a plant considers a bite by an animal to be a wound, which requires a healing process analogous to what happens when a scab forms on your body. The plant needs nutrients from the soil to complete the healing process and sets to work collecting those nutrients by releasing more root exudates to attract and feed carbon-hungry microbes.

According to Dr. Nichols, this level of stress is good for a plant, which otherwise tends to be "lazy" and not work as hard as it could for nutrients. The scientific term is the *conservation of resources*—which means no organism will produce anything more than what it needs. Plants seek balance and dynamic equilibrium, and they need stress— but not too much—to reach peak performance. Stressing plants is like training for the Olympics. Your body won't be ready unless you stress it in the right ways to get in shape. For a plant to acquire extra nutrients, it needs to work for them. This is why livestock are a key component of regenerative agriculture and soil health. In the conventional farming model, plants don't work for their nutrients—they get them from us at great expense to our pocketbooks!

Three

Regenerative Revelations

Mistakes and failures are inevitable in farming, and the silver lining is the lessons we learn from them. But learning through experiments and crop trials—such as our paddock water supply system and planting diverse mixes of covers crops—was a lot more fun than learning from failures! By 2007, the results from the four years of split-crop trials with and without synthetic fertilizer proved without a doubt that Dr. Kris Nichols was right. For four years in a row, the crop yields of the unfertilized half of the test fields were *equal to or greater than* the fertilized half! I also noticed a dramatic improvement in the health of our soils once I removed synthetic fertilizer. The soil was much more aggregated. It really did look like chocolate cake! This aggregation meant water infiltration had improved significantly. (See plate 9 on page 4.)

I probably don't need to tell you that we haven't used any synthetic fertilizer on our owned land since that time. We discontinued it on rented land in 2010. How did the decision to stop using fertilizer affect our production? Our yields today are approximately 20 percent higher than the average in the county. Do I have the highest yields in the county? No. Am I one of the most profitable operations in the county? I think so. One reason, of course, is straightforward: I am no longer spending money on synthetic fertilizer, pesticides, or fungicides. Simple as that. But as I noted above, I am *not* advising you to immediately eliminate synthetic fertilizer on your operation. You're going to have a disaster if you do. Soils accustomed to synthetic fertilizer are like drug addicts: They need to be weaned off their addiction slowly. It's essential to first restore your soil's ability to function by encouraging the growth of living plants to feed soil biology. Seeding multispecies cover crops—and, preferably, having livestock graze those crops—is an excellent way to do this. Once soil biology diversifies and proliferates, including healthy populations of mycorrhizal fungi, you will be able to drastically reduce your synthetic fertilizer inputs.

Our decision in 2007 to stop using synthetic fertilizers on the ranch was a big step on our journey into regenerative agriculture. It was easy at that point to see the difference between our soil and our farming neighbors, including an organic producer. Their croplands clearly lacked organic matter and soil structure. The water infiltration rate on our farm had increased significantly. In 1991, the rate was one-half inch per hour. In 2015, it was one inch of water in nine seconds. A second inch will infiltrate in sixteen seconds. That's two inches in twenty-five seconds! That's the power of mycorrhizal fungi and soil biology. They combine to build soil aggregates, which allows water to infiltrate, and then the organic matter stores that water. It's not a question of how much total rain falls on your land, it's how much can infiltrate into your soils and then be stored there that counts. That storage ability is called *effective rainfall*. If we have low amounts of effective rainfall, we create our own drought.

The Critical Role of Carbon

In my ongoing reading and research about soil and plants, one thing that kept coming up was the importance of carbon in the system. When I came across the website amazingcarbon.com, I was fascinated. Dr. Christine Jones, a soil ecologist in Australia, developed the site to help others understand the critical role carbon plays in ecosystem function, particularly underground. Dr. Jones clearly explains how soil carbon is the key driver for much of soil health. Soil carbon is also critical to water-holding capacity. Thus, she concludes, soil carbon is the key driver for farm profit.

So how do we get more carbon into our soils? Imagine for a moment a stretch of agricultural land. As temperatures warm in the spring and the sun rises higher in the sky, seeds planted in the soil of this land germinate, and the seedlings form roots that fan out underground in search of the water and nutrients necessary for survival. According to Dr. Jones, "plants take in carbon dioxide from the air and combine it with water to form simple sugars. These simple sugars, referred to as photosynthate, are the building blocks of life. Plants transform these sugars into a wide array of other compounds. Many of these compounds are used by plants for growth, however, a significant amount of them are transferred to the root tips where they are 'leaked' into the soil as root exudates."

Why would a plant leak those exudates, which Dr. Jones called "liquid carbon," into the soil? To feed microbes, of course! A myriad of life forms use liquid carbon as a food source. The plants, in return, benefit from the nutrients released from the soil and transferred to their roots. Consider that 95 percent of life on land resides in the soil, and you'll realize just how important this relationship is. Add to this the fact that, as Dr. Jones explained on her website, "microbial activity also drives the process of soil aggregation, enhancing soil structural stability, aeration, infiltration, and water-holding capacity. All living things, above and below ground, benefit when the plant-microbe bridge is functioning effectively."

Some forms of organic carbon, such as crop residues, animal manures, or compost, can be spread on the soil surface. These visible materials (called organic matter) have many physical benefits, but they eventually decompose, producing carbon dioxide. Root exudates, on the other hand, are key to soil building, because they are the main source of carbon for microbial communities deep in the soil profile. Microbes supported by root exudates are essential to the production of humus, a highly stable and long-lived form of

The Rhizosphere

The top layer of a typical soil profile is called the A horizon, where a vast array of soil life can be found, including bacteria, fungi, protozoa, nematodes, and earthworms. Varying from two to eighteen inches deep, or deeper, the A horizon is where much of the liquid carbon ends up when it is exuded from the roots of plants. Microbes use some of the energy from the liquid carbon to release and transfer plant nutrients and some of it to create stable carbon compounds essential to the formation of *topsoil*.

Extending another twenty inches or so below the A horizon is the B horizon, which is often called the *subsoil* layer. It consists of minerals and a lesser amount, compared to topsoil, of organic carbon and biological activity—at least until plant roots penetrate it, at which point the microbes follow. Over time, this process will transform the B horizon into an extension of the A horizon.

The next layer in the soil profile is the C horizon, which consists of rock and other unconsolidated parent material that has not yet been weathered or decomposed into smaller particles. This layer provides the source material for B and A horizons. When we talk about topsoil eroding into the sea as

organic carbon with high cation-exchange and high water-holding capacity. When soil is enhanced at depth, the function of the entire watershed is improved, with benefits that extend to freshwater and marine environments far from the farm.

In healthy, living soils covered with green plants for much of the year, the carbon supply for beneficial soil microbes can be nearly endless. I cannot emphasize this enough: This process is absolutely key! According to Dr. Jones, the formation of fertile topsoil can be a consequence of industrial practices, it is the vibrant, vital A horizon that is being lost. When we talk about restoring degraded land to health, it primarily means the creation of new topsoil from subsoil, a process that can happen much faster than we used to think possible.

The primary agents for converting one horizon into another, thus creating the right conditions for agriculture, are microbes and plant roots. The soil–root interface is called the *rhizosphere*, a name coined in the early twentieth century by Lorenz Hiltner, a pioneering German soil scientist who studied the effects of beneficial microbes on plant health and nutrition. The rhizosphere is the zone surrounding a plant root on all sides, often only a few millimeters wide, and the scene of highly concentrated biological activity. Hiltner determined that soils with high densities of microbes conveyed significant health advantages to plants over soils with diminished microbial life. This discovery ran counter to the prevailing attitude at the time among scientists and other professionals that "the only good microbe is a dead microbe."

breathtakingly rapid once the biological dots have been joined. The sun's energy, captured in photosynthesis and channeled from above ground to below ground as liquid carbon, fuels the microbes that solubilize minerals. A portion of the newly released minerals enable rapid humification in deep layers of soil, while others are returned to plant leaves, facilitating an elevated rate of photosynthesis and increased production of plant sugars. This positive feedback loop makes soil-building somewhat akin to perpetual motion.

For a long time, scientists believed that plant roots released microbe-attracting exudates passively. But, as it turns out, plants are every bit as calculating as animals in securing the resources they need to survive. Call it plant intelligence. To gain biologically available access to a needed nutrient, a plant must attract specific microbes that are genetically hard-wired to solubilize that particular mineral. The process is not yet fully understood, but it goes something like this: A plant sends out a chemical signal via its exudates that it needs a particular nutrient, such as phosphorus, and the microbes attuned to this signal respond accordingly. If a plant has an additional nutritional need, it generates a different signal, catching the attention of still other microbes. As you can imagine, the communication gets complicated quickly. However, the beauty is that natural ecosystems have figured this out. One plant can signal for one nutrient and an adjacent plant can signal for another, and the system responds perfectly.

Dr. Christine Jones explains how applying too much nitrogen can suppress the association that microbes have with plants. The plants and microbes will use the nitrogen independent of each other, thus delaying the vital associations between them. Later in the growing season, when the plant needs the microbes in order to supply critical nutrients, the plant will not be able to access them. This leads to lower yields. After I learned this, I better understood why, when I ran the four years of fertilization trials on my cropland, I saw the results I did. By not fertilizing, I was actually encouraging this natural plant–microbe association.

In a healthy rhizosphere, microbes and plant roots quickly establish their two-way communication process. The quantity and variety of this "call and response" is staggering. And we still have a huge amount to learn about the rhizosphere. Scientists say that 90 percent of the planet's estimated *one trillion* microbial species have yet to be discovered. A recent study involving a research technique called metagenomics added twenty new bacterial phyla to Earth's tree of life. To put this in context, all insects on the planet belong to a single phylum (as do all animals with backbones); perhaps that gives a sense of how big this microbial universe might be. Add in all the signaling going back and forth among soil microbial species that we don't even know exist yet, and you can see why this underground universe, billions of years in the making, has become a vast frontier of scientific investigation.

The Fusion of Life

We began hosting many farmers and others on summer tours of our ranch, and as I talked with these groups of people, I always enjoyed educating them about the incredible unseen interactions between plant roots and soil microbes. I first met Ray Archuleta when he came on one of those tours. At that time Ray worked for the USDA's National Technology Center in North Carolina, part of the Natural Resources Conservation Service (NRCS), where he trained people from across the country in improving soil quality. Little did I know then that we would soon become not only best friends but business partners as well.

One thing I remember well was the puzzled look on Ray's face during the tour as I explained my soil-management methods. I assumed he was disagreeing with the things I was saying, but, as he told me later, his expression reflected a revelation about soil he was having. I like to tell people that it was the moment the wiring in Ray's brain crossed—leading to big changes in his thinking and his work. Ray soon became known as "Ray the Soil Guy!"

Ray began reading books by Allan Savory, as well as any publications on soil health and microbiology that he could get his hands on. He called it the first step in his "deprogramming" from nearly everything he had been taught about agriculture up to that point. What Ray heard and saw during the tour of my ranch was exactly what he had been searching for—and this dawning realization was what I could see unfolding through the look on his face. Although he had spent nearly eight years studying agronomy at universities, his professors had not taught him a thing about how the soil functioned, especially not about soil biology. He had been taught—as I had—that agriculture was all about killing living things with pesticides, insecticides, and fungicides and applying chemical inputs to drive crop growth. But when he saw our eight-way and ten-way cover crop mixes alive and well in a dry year, he realized that healthy soil is a living ecosystem. "Health is life and life is health," is how he describes it today. He realized that nature is more collaborative than competitive, and this concept ran totally against his training as a government agronomist.

Ray came up with a catchphrase as a result of his conversion: He called what's happening in healthy soil the "fusion of life." Geology is sand, silt, and clay—dirt, in other words. The fusion of life transforms dirt into soil. Dirt becomes soil not simply because there is enough organic matter in the soil but because there's life in the soil—and not just any life but the full spectrum of soil biology. As Ray likes to say, without life we might as well be farming on the moon.

Microbes can replicate astonishingly quickly, which means nature can easily be self-healing and self-regulating if we give her a chance. But too many modern agricultural practices destroy the self-healing process, with tillage as the primary culprit. Once farmers stop tilling, they can further the healing process by planting a cover crop because, as Ray has pointed out, green plants don't just protect the soil, they are *biological primers*. They capture solar energy and transfer it to the microbes in the soil, fueling the fusion of life. Without cover crops, you're going to "spill the sun," as Ray says, and waste an opportunity to boost the healing process.

One of the biggest challenges we face in the twenty-first century is the growing disconnect between people and the land. We see this disconnect not just in young people in cities but in farmers and ranchers, too. Very few people understand that the soil is an ecosystem, so it is our duty to educate as many people as we can that the soil is alive. After the tour on Brown's Ranch, Ray developed a passion for soil biology and began to give talks all around the country, which ignited the soil-health movement.

Although his job with the NRCS was to improve soil quality on ranches and farms, Ray decided the agency was failing in its mission. Everywhere he traveled, he saw topsoil eroding into rivers and streams despite the billions of dollars spent on conservation practices by government agencies and landowners over the decades. One of Ray's favorite sayings is "our lakes and rivers are filled with conservation plans and nutrient plans but not crystal clear with understanding!" This saying made some NRCS officials quite upset. But they were missing his point: Nutrient and conservation plans can be helpful to landowners, but they should not be the goal of the work. Understanding how the soil actually functions is the goal! Ray could see that, unfortunately, this message wasn't getting through at NRCS. According to the Environmental Protection Agency, for example, despite all the conservation and nutrient plans written by NRCS over the years, sediment (i.e., eroded soil) is still the number one water-quality problem in the country.

The clincher for Ray, however, was when a farmer friend of his admitted that he couldn't bring his son into the business because there wasn't enough profit for two families. That's when Ray really started to question the modern agriculture model. (The issue of bringing the next generation into agriculture is critical, and I discuss it in chapter 5.)

It's a miracle that Ray wasn't fired from his job with the USDA. One of the messages that Ray repeated in every one of his presentations was about what happens when we destroy soil structure. He used a slake test, consisting of four tall, clear plastic tubes filled with water, into which he invited participants to drop different clods of soil representing different types of farming: conventional and regenerative. The clods from the high-tillage, conventional farms dissolved almost immediately, demonstrating how poorly the soils were bound together. The no-till and regenerative clods, in contrast, held their shape in the water for a long time, signaling their structural integrity. That's why it was called a slake test—when water rushes into the millions of microscopic pores in a clod that has less structural integrity, the clod will begin to slake, or break apart. This disintegration is an indication that the biological glues that hold the soil together are weak or absent. Slaking reduces infiltration rates and increases the risk of soil erosion. Ray also had a rainfall simulator in which he compared how soil under different management practices handled a rainfall event. As you can imagine, the soils from regenerative farms performed at higher levels in the rainfall simulator as well. These tests were Ray's way of starting a conversation about building locally robust, productive soils using cover-cropping strategies.

Diving Deeper into Crop Diversity

A few years after I first met Ray, he and I took a trip to David and Kendra Brandt's farm near Carroll, Ohio, to speak at a field day. Ray had been to the Brandt farm many times before, but it was my first visit. We arrived a day early in order to have time to tour the farm. As we pulled up to the shop, David emerged clad in his favorite attire, a pair of blue coveralls. At our first meeting, David was an intimidating man. Standing six feet three inches tall, he greeted us in a loud, booming voice and extended a hand the size of a catcher's mitt. David is a matter-of-fact type of person who sticks to his beliefs and is not afraid to tell you what he thinks and why.

His story is compelling. Few people have his experience and expertise. Like most other producers in that area, David grows corn and soybeans. Unlike other producers, he has been practicing no-till since 1971 and using cover crops since 1978, going against the norm all of his adult life. Many years ago, David added winter wheat to his crop rotation. He told us that he did this to not only diversify his cash crops but in order to give him the window of time he needed to plant a cover crop to address soil-health concerns.

After we exchanged greetings, David took Ray and me to a field in which knee-high field peas and daikon radish were growing. As we walked into the field, David proudly explained how many pounds of nitrogen the peas in the field were producing via rhizobia and how the radish scavenged the nitrogen, stored it, and then released it the following spring as the tubers decayed. I glanced at Ray, and he was about ready to bust out laughing. He knew that I was having a hard time keeping my mouth shut. Finally, I burst out with, "David, why only two species?" David looked at me, shocked that I would ask him that. He had expected me to be impressed, and there I was questioning him. "Now Gabe," he said, "we can't grow those multispecies cover crops here!" I pointed out that if he would go visit some of the little bit of native rangeland left in Ohio, he would see diversity does work there. Ray, meanwhile, had turned away so David would not see him laughing.

My bold statement set David to thinking. To his credit, he took the challenge and started planting diverse cover crops, doubtless hoping to prove me wrong. Today, though, he is a real believer. Such a believer that he has spent countless hours working with Dr. Rafiq Islam of Ohio State University and Jim Hoorman of NRCS to quantify the positive effects of cover crops. Year in and year out David grows 200-plus bushels of corn per acre with little to no synthetic fertilizer. The biomass of cover crops David and Kendra grow makes my mouth water thinking of how much livestock I could graze on it.

David's soils are absolutely amazing. You can stick a spade into any of David's fields and reveal eighteen-plus inches of dark chocolate cake–like topsoil. If you step over onto neighboring property that has been farmed conventionally with tillage, and you stick in your spade, you will find tight, yellow clay. The contrast is stark. It is a true testament to the power of regenerative agriculture.

The year after I first visited David's ranch, I met Dr. Jonathan Lundgren when he spoke at the No-Till on the Plains conference. Dr. Lundgren worked for the USDA's Agricultural Research Service at that time, and I was ecstatic to hear an entomologist talk about beneficial insects, not pests. His talk reinforced for me the importance of thinking of my farm as an ecosystem, and in so doing, providing a home for beneficial insects, both predators and pollinators.

According to Dr. Lundgren, there are between 3,500 and 15,000 insect species worldwide that can be considered pests, in terms of human endeavors. These species eat our food, destroy our homes, bite our children, and transmit diseases. In fact, over the centuries insects have killed more people than died in wars! Most people have a negative attitude about all insects, just as they do toward "germs" and bacteria. But for every pest species, there are between 400 and 1,700 species of insects that are *beneficial* to humans. Without these beneficial insects, food webs and ecosystems would collapse. Humans depend on insects. If you like fruits, vegetables, or flowers, then you can thank a bee, beetle, or butterfly. Many cultures around the world eat insects as part of their diet. And insects are the food source of many other species, including many that are important to us—birds, for example. Soil invertebrates, which include insects and earthworms, are also critical to soil health. New research shows there could be as many as *one billion* soil invertebrates per acre in healthy soil.

Dr. Lundgren believes insects are nature's pesticides—the good ones eat the bad ones. Some predator insects are amazingly effective. For example, lady beetles can decimate pests. If a farmer has an infestation of pests, it is because there is a lack of predatory insects. Most farmers use insecticides to kill pests, but what they don't realize is that they are killing predator insects, too. Thus, they ensure that they will never have the population of predatory insects they need to kill the pests. The key to having a healthy population of beneficial insects on your farm is diversity and not using insecticides, according to Dr. Lundgren. Regenerative practices encourage biodiversity, which combats pests. Science backs this up.

"Increasing diversity in simplified systems is associated with fewer pests," he told me, "and we have documented that the balance of species networks were predictive of pest abundance in corn fields on actual farms. Specifically, more diversity and a greater balance of species' abundances within a community of insects leads to fewer pests."

From a scientific perspective, Dr. Lundgren believes this has something to do with the astounding complexity of insect–plant relationships. When there is a lot of plant diversity in a field, one of three things may be going on:

1. There are too many plants for pests to have a big impact.
2. There is increased competition from beneficial insects.
3. The plant itself changes physiologically as a result of what's happening in the soil, especially as it improves its health from a degraded and poisoned condition. Or the plant responds to the presence of a beneficial insect, and that makes it less attractive to pests.

The simplest thing to do to help beneficial insects, Dr. Lundgren said, is to stop tilling. Then put in a cover crop with as many species as possible. Notice that these are the same two practices that Ray Archuleta recommended as the way to start healing our soils. Tillage significantly reduces the biotic resistance of plants to pest proliferation. Bare soil is bad, too.

When I talked with Dr. Lundgren at the no-till conference, I was thinking about bees and dung beetles but not many other kinds of insects. I didn't appreciate the full range of services that insects provide. I have since had the opportunity to host Dr. Lundgren at my ranch several times, learning more each time. And he has shared with me that working with me and Paul and other regenerative farmers has also taught him that he needed to rethink how he did his science. Instead of doing a traditional replication experiment, gathering all the data then publishing it later in a peer-reviewed journal, he decided instead to try to understand what was working on our farms. So he came out with his team of research assistants and graduate students, and they counted as many of the insects as they could in a given area. He also did comparisons between conventional versus regenerative operations in the area. He found that regenerative farms had ten times fewer pests than conventional farms! This work has now been peer-reviewed and published.

His recent research, which he calls practice-based science, indicates there is no such thing as too much diversity. It also suggests that beneficial insect populations can recover quickly after the adoption of regenerative practices, even within one year.

After learning from Dr. Lundgren just how crucial it is to have a diverse population of insects on a farm, ranch, or garden, we set out to increase them in our operation. We took a look at a map of our land and marked fields and locations where we could seed what we call pollinator strips. These pollinator strips are comprised of annual and perennial grasses, forbs, and legumes whose main purpose is to provide a home for pollinator and predator insects. We seeded species such as clovers, chicory, warm-season grasses, plantain, birdsfoot trefoil, coneflowers, and others. (See plate 18 on page 9.) We also seeded this combination in our orchards. These strips can be grazed by our livestock on occasion, to generate even more dollars. Wildlife also thrive in these strips. Nearly every farm or ranch has an odd-shaped field that could be turned into a pollinator strip. Why not?

The "Chaos" Garden

David Brandt and another farmer friend, Gail Fuller of Kansas, and I like to try to outdo each other by seeing who can come up with the craziest "experiment" to regenerate soils. In 2012, with Dr. Lundgren's talk in mind, I set out to find out how a very, *very* diverse cover crop would perform.

I started with over twenty cover crop species, including pearl millet, proso millet, German (foxtail) millet, cowpeas, crimson clover, soybeans, arrowleaf clover, berseem clover, sunn hemp, buckwheat, flax, oats, lentils, and sunflowers. To this I added over twenty species of annual flowers: calendula, asters, begonias, daisies, cosmos, geraniums, marigolds, pansies, morning glories, petunias, snapdragons, and more. The main component of the mix, however, was vegetables: five varieties of sweet corn, four varieties of peas, four varieties of beans, multiple varieties of squash, watermelon, muskmelon, radish, turnips, carrots, lettuce, spinach, eggplant, tomatoes, tomatillos, zucchini, kale, beets, cabbage, cauliflower, onions, and more.

Altogether, I planted a 30-acre "garden" with this mix of over seventy species of vegetables, flowers, and cover crops. Even though precipitation that year was well below average, the mix thrived. Which was not surprising, really. Remember the cover crop cocktail demonstration on the Burleigh County Soil Conservation District's plot land that I described in chapter 2? (See *Cover Crop Cocktails*, page 32.) Bacteria, protozoa, fungi, nematodes, earthworms, pollinators, predators, and the entire soil food web of life was feasting and thriving in my 30 acres! Everyone who saw this garden was amazed by its productivity. It was chaos; hence, it was named the "Chaos Garden"!

It was incredible to see the diversity of insect species in that garden as well. They set me to thinking about what Dr. Lundgren said about their importance in a healthy ecosystem. Whenever I show visitors around the ranch, I like to take them to our home garden first. Everything one needs to know about a how an ecosystem functions can be seen, touched, and smelled in a healthy garden.

Although that experiment was fun, a chaos garden is not practical. First, it is too costly. Second, it is difficult to harvest. Shelly would send me to pick a vegetable for supper and whatever I tripped over is what I would bring home. I stepped on a lot of vegetables trying to harvest them!

The experience had ramifications on how we have planted our garden since then, however. Instead of a tangled biomass of mixed-up plants, we plant a row of sweet corn, for example, and fifteen inches to one side we plant a row of peas, fifteen inches to the other side a row of green beans. It's a little ecosystem, with the grass plant (corn) cycling phosphorus, and legumes (peas and green beans) cycling nitrogen, being transferred to each other by mycorrhizal fungi. Does this sound familiar? We are easily able to hand-harvest the individual rows this way and yet have the diversity of species that a healthy ecosystem requires.

We tried another successful experiment in our main garden as well. The garden was fenced, to help keep the deer out, and measured approximately 150 feet by 150 feet, to which we added two *hugelkultur* beds. What is *hugelkultur*? The word means "hill culture" in German and is a technique where woody debris is utilized as a resource in farming or gardening. In our case, we took logs cut from dead trees found on our ranch and laid them out as a frame that was 12 feet wide by 100 feet long. Then, we put logs and branches inside this frame along with a mix of compost and soil. This formed a mound that was about 4 feet high—hence, a hugelkultur. More wood and compost was added on top of the mound as it decayed over time and turned to soil.

Visitors to the ranch often ask why we put all this wood in our garden. The answer ties back to the most important element on a farm: carbon. Wood is high in carbon, which becomes food for biology. It also holds moisture, which is a real benefit in a dry environment. Over the years, the wood in the hugelkulturs has broken down via biology, creating a very healthy soil in the garden, which leads to nutrient-dense vegetables. (Chapter 10 explains the importance of nutrient-dense food for human health).

The remainder of the garden is, of course, not tilled—even our potatoes. To "plant" potatoes we simply place the seed potatoes on the soil surface and then cover them with a thin layer of second-cutting alfalfa hay, but not too thick, otherwise the tender shoots have a hard time growing up through the hay. As the hay breaks down, consumed by biology over the course of a summer, we simply put more hay around the plants to keep weeds from germinating. When we want to harvest potatoes, we peel back the hay and viola—potatoes! (See plates 20 and 21 on page 10.) No digging and easy to clean. The potatoes do tend to be smaller in size when grown with this method. Perfect for grilling, I say.

As I described, the rest of our garden is planted in individual rows with each row a different vegetable or flower. Don't forget the flowers to attract the pollinators and predators. By planting plenty of flowers we also have the option to sell some of them as bouquets at the farmers market, another income stream.

In late fall, after all of the vegetables are harvested, we pull our mobile chicken coops (which I describe in chapter 5) onto the garden, letting the chickens clean up any unpicked vegetables and greens. Their droppings add a nice layer of natural fertilizer.

Speaking of chickens in gardens, I once participated in a radio interview with a program based out of California. The topic to be discussed was how to regenerate soils on vegetable farms and gardens. I called in from my home in North Dakota. Also on the program was a soil scientist from California. The host asked me to start by telling the listeners about our garden. I proceeded to tell them how we focus on producing nutrient-dense, great-tasting food by no-tilling diverse mixes of vegetables, along with grasses, forbs, and flowers, to attract pollinators and predators. Then I explained how we run our chickens on the garden in the fall to help fertilize and stimulate soil biology. After a commercial break the host asked the soil scientist what she thought of my approach. "Well, we can't do that here!" she exclaimed. "There is no way you should let a chicken near a garden! It may contaminate the vegetables. In fact," she continued, "we fumigate all of the vegetables just prior to harvest to ensure that there isn't even an insect alive when we harvest the vegetables!" I just couldn't help myself. I interrupted, asking, "Are those the same vegetables that you want our children to eat?" The host abruptly interrupted me by exclaiming, "We'll take a break now!" I had to laugh, but how sad it is that we are more worried about a chicken eating insects in a garden than we are about spraying our vegetables with insecticides.

Back to our garden. Our chickens do a great job, and after they're done we unroll a round bale of second-cut alfalfa hay over the entire garden. We do this for a number of reasons. In our environment there are not enough frost-free days to grow a cover crop following the vegetables, so the alfalfa hay provides the armor that is needed to protect the soil and to provide food for macrobiology. In the spring, we "part" the hay and seed our vegetables. Over time, the alfalfa, which is high in nitrogen, is consumed by biology, and then wood chips are added as mulch to prevent weeds. Wood chips and alfalfa provide a balance of carbon to nitrogen. (I will explain the importance of the carbon:nitrogen ratio in chapter 7, where I delve into the all-important principles of soil health.)

Just as we do with our grain crops, we harvest seeds from many of the crops we grow in our vegetable gardens. Our growing season isn't long enough for some of the species to set seed, but many do. I believe it is a real advantage if you can save seeds from plants grown on your own place, be it a farm or garden. If a plant can grow to maturity and set seed, that's pretty good evidence that it is a healthy plant. And that's the goal, isn't it? Healthy plants, healthy soil, healthy ecosystem.

Four

Rethinking Our Livestock Focus

For several years after the disaster years, I continued to focus on animal performance in the management of our cattle herd. But as the years went by, I started to realized that some of my livestock management practices could be made more holistic, too. Because I had been focusing on animal performance, the mature size of our cows had grown ridiculously large. By 2007, they averaged over 1,400 pounds! It was costing way too much money to feed those animals. I noticed that the few small mature cows we had left were always in good condition and they always bred back. Observing this led me to an important change in my thinking (which, as I've already pointed out, is more important than the "doing"): The size of our cows no longer matched the environment. They were too big! For twenty-six years I had been raising and selling registered bulls. I touted *numbers*—weaning weights, yearling weights, or EPDs. I came to the realization that those numbers were basically meaningless when it came to determining profitability. What mattered was having cows that could convert forages to meat on my operation. The focus of the production model I was using—on continually increasing pounds—had led us down the wrong path. We needed to focus on *profit per acre*, not pounds of animal produced.

We began to select bulls and replacement heifers born from smaller cows that had been in the herd for at least four years. We started

Thoughts on the
Registered Cattle Business

For well over twenty years, I registered my cattle. A registered animal must meet several criteria. These criteria vary a bit from one breed association to another but generally they include:

- The animal must have a permanent tattoo for identification.
- The animal's sire and dam must both be registered.
- Birth weights, weaning weights, and yearling weights must be taken and reported to the appropriate breed association.
- The breed associations collect the data and use them to develop expected progeny differences (EPDs). EPDs are a projection of how an individual and his or her offspring will perform. EPDs are developed for birth weights, weaning weights, yearling weights, calving ease, milking ability, carcass traits, and many other characteristics.

breeding this herd to bulls with smaller frame scores. This helped us bring the frame size down and move our herd toward the type of cattle that could graze longer throughout the year and required less "groceries" to keep them going. Reading Walt Davis's book *How to Not Go Broke Ranching* and Chip Hines's book *How Did We Get It So Wrong* taught me a lot about the fallacies of the traditional beef production model. I just wish I had read their books early in my ranching career.

Along with the downsizing of mature cow size came other changes in management. I have never butchered a beef animal

In college I was taught that in order to "improve" my cattle it was beneficial for me to buy and use only registered bulls. There is no doubt that by registering cattle and studying EPD's one can focus on improving individual animal traits. The problem is that the livestock industry has focused solely on the traits of individual animal performance. This has led to larger and larger mature cattle size, which, although usually good for feedlots and packers, leads to cow and calf producers having a cowherd whose mature size is too large for their environment. This leads to decreased profitability.

For over twenty years, I followed that mantra: Use only registered bulls. I spent tens of thousands of dollars registering cattle and selling my stock to cow/calf producers. Today, looking back, I realize how foolish I was. I have learned that I wasn't really improving the bottom line for my customers. Smaller mature cows allow an operator to run more animals on a given acreage, compared to larger cows, which means smaller cows will always give a higher net return per acre.

and found a gizzard inside. So I asked myself, *Why am I feeding these animals grain?* That is not how ruminants evolved. We were already raising some grass-finished beef for our own consumption, due to the health benefits of eating grass-fed meat, so why feed the remainder of our herd grain? This realization led to a major change in our business. Our February 2009 bull sale was our last. Our customers were puzzled when we informed them that we were getting out of the bull business. They did not understand that it did not meet our holistic goals, one of which was farming and ranching in nature's image.

We also decided to stop using wormers, fly-tags, and the long list of vaccines. These products were Band-aids treating symptoms. They were not solving the real problem, which was a dysfunctional ecosystem. That summer we waited until July to turn out the bulls with the cows, so that the cows would calve in April instead of the extremely cold weather of February and March. Instead of maintaining six separate herds, we reduced to three (each with multiple bulls), and we shortened the breeding season to sixty days. We also started moving the herds more frequently, thanks to our newly built grazing system. This allowed us to run higher stock densities while allowing longer recovery periods.

In 2010, we pushed our breeding season back even further; we waited until the first week of August to turn out the bulls. The cows were exposed to the bulls for only forty-five days. We combined all the animals into one herd, which allowed us to address our resource concerns even further.

When the 2011 calving season arrived, I knew we were finally in sync with nature. We were calving during the time of year when the deer were having their fawns. By changing our calving date, we no longer had to worry about blizzards, mud, ice, sick calves, dirty udders, frozen ears, confinement, stressed cattle, stressed people, bedding corrals, babysitting first-calf heifers, and bragging about how hard we work. The cows calved in a nice, clean environment on a high level of nutrition and the calves were very healthy. Making these changes were some of the best management decisions we have ever made on our ranch.

Allowing a Cow to Be a Cow

Our calving management schedule continues to serve us well. During calving, Paul moves the cows daily. Any newborn bull calves are banded (castrated) at birth, except for bulls born to an old cow who has a good udder, feet, and legs and is easy fleshing. Those calves are left to mature as bulls for use in our own herd. What better place to find bulls for our operation than from our own operation? We have found this is an excellent way to build an efficient, profitable cowherd.

During the winter months of December, January, and February, we prefer to graze the cow/calf pairs on cover crops. My cover crops of choice for this are: brown mid-rib sorghum/sudangrass, along with hairy vetch (which will still be around 18 percent crude protein in winter), kale, and collards or some other forage brassica. The remaining forage comprises species that address whatever resource concerns I have at the time. Annual ryegrass makes for good late-fall and winter grazing, as does hairy vetch. For more about what I feed the animals, see *Managing with Flexibility* on page 69.

Once the cover crop freezes and winter sets in, we do not move the cow/calf pairs daily. Yes, we would get better utilization if we did, but because I travel to speaking engagements from October through March, there is less on-farm labor available. When managing the ranch on his own, Paul does not have time to move the animals more than every few days. It's not a perfect world! And remember, part of regenerative agriculture is regenerating our own minds and bodies. So don't be afraid to take a break and ease your workload a bit.

I am often asked what type of fencing we use when we graze cropland. We have taken the time and expense to put permanent high-tensile electric fence around the perimeter of all of our owned and rented properties. This allows us the security of a permanent fence and the ability to transfer power to temporary fences used on the cropland.

We have shallow water pipelines buried throughout our ranch, as described in chapter 2. Risers were plumbed in at various locations, allowing easy access to water. We set a rubber tire tank near the riser, hook up a garden hose and a float, and we have water. We prefer to put the risers near the middle of the field, not on the edge, because this allows us to graze first one direction from the tank and then the other.

Starting near a water tank, we string a polywire through tread-in ring top posts across the field, connecting it to the permanent high-tensile fence. We leave this polywire in place. We set up another temporary fence at an appropriate distance to provide the livestock one days' worth of grazing, allowing them access to the water. Within that temporary fence is the first paddock grazed. After the livestock have grazed to the desired amount, we string another fence to provide one day's worth of grazing, further away from the tank. We then roll up the previous days' fence, allowing access to the new forage. And so it goes, until we reach the end of the field. The cattle must walk back over ground they have previously grazed to reach water. But because we use high stock densities, it is only a matter of days before we finish grazing that half of the field, so there's never a problem of soil being beaten bare by animal traffic to the water tank.

Once we reach the end of the field, we proceed to make paddocks in the opposite direction from the water tank. We set up a back fence so the cattle don't walk back over the half of the field that was previously grazed.

What about rainfall events? Do I take the cattle off the cropland? No, I do not. They have to be kept somewhere, and I would rather have them continue on the diet they are on, rather than moving them to a perennial pasture and changing rations. Does keeping the animals on wet ground cause much damage to the cropland? No, it doesn't on my land. During heavy rainfall events, some pugging may occur, but after a year or two it tends to smooth back out. In heavy clay soils, of course, pugging may be more extensive. My advice is: Whatever happens, don't get discouraged and resort to using a tillage implement. Just relax and observe. Nature will take care of it. In the grand scheme of things, a small amount of pugged acreage is not going to break you.

A More Natural Way of Weaning

One important point about our management is that we do not wean our calves in the fall. The heifer calves need to learn how to become

Epigenetics

Epigenetics is the study of heritable changes in gene function that do not involve changes to the DNA sequence. It is the study of biological mechanisms that will "switch" genes on and off. For instance, what an animal—or a human, for that matter—eats, where it lives, how it is handled, what stresses it encounters in life, and how it ages are all factors that can cause chemical modifications at a cellular level that can, over time, turn genes in those cells on or off.

Hence, many experiences that an animal has during its lifetime may have consequences for future generations. This is one of the reasons why we leave the calves with the cows outdoors on pasture all winter. It is our belief that, by consuming lower-quality forages, those calves will develop the capacity to thrive on lower-quality forages throughout their lives. The fact that our cows are feeding on lower-quality forage while their calves are in utero also contributes to the calves' ability to adapt. Taking advantage of this epigenetic phenomenon significantly improves the potential for our animals to return us a profit.

cows that can thrive in our environment. As the calves graze alongside their mothers all winter, they learn which plants to eat and which to avoid. They also learn from their mothers how to tell if a storm is approaching and how to trail back to the farmyard for protection. They learn how to use snow as their water source. (We do allow them to access water from a winter water supply in the farmyard, but most of our cows will not travel for water as long as snow is available.)

In early April, we fence-line wean the calves. This is a simple process of splitting the calves away from their mothers and keeping them apart with an electric fence separating them. The calves can see the cows, they can even touch noses, but the fence prevents them from nursing. The calf is content, the cow is content, life is good. We set up this arrangement in a way that requires the cows to walk a distance away from the fence in order to graze. After one or two trips back to check their calf, they get tired of walking and just stay out grazing. Four or five days later, we move the calves out to a paddock where we stockpiled forage during the previous growing season. The calves are used to grazing this type of forage, and they take right off. We find that the calves stay healthy using this strategy. Weaned calves, the easy way!

The calves graze on perennial pastures and are moved once a day until early August. At that time, we separate the steers from the heifers, and the heifers are exposed to bulls for thirty days. After thirty days, the bulls are pulled, and the steers are put back with the heifers. This allows us to graze at higher stock density and also lessens the workload. People often ask why we don't run all of the cattle— cows, calves, and yearlings—together. From an ecological perspective it would be better to do so, but due to the fact that our land is not contiguous, it would be too time consuming and laborious to load them all up into trailers and move them several times during the growing season.

In early December we determine if the heifers are pregnant through ultrasound. Those that are pregnant are grouped with the mature cows. Those that are not go on to be grass-finished. We do not check the mature cows for pregnancy. What advantage would it serve? Even if some cows were not pregnant, we would not choose to cull them at that time, because then we would have to take care of the weaned calves. Extra work like that is not what we are looking for. Instead, we run all the cows, open or not, with their calves on them throughout the winter. After weaning time, the cows graze on fresh new grass growth. They flesh up well, and any open cows really get fat. In late June we set up some portable panels in the pasture and pull any cow that does not have a calf nursing on her. She was open or she lost a calf at birth, but either way, it's a sign that we should not keep her in the herd. Those cows are hog fat, and when is the hamburger market booming? Right around the Fourth of July, of course! We save the expense of pregnancy testing, and we have excellent hamburger meat to market at a more lucrative time of the year instead of selling open cows in December when prices are low.

We never give open animals a second chance. They are sold, period. We select for animals that can perform in our environment, which helps to ensure profitability.

We like to keep our cowherd constant at around three hundred head. Yearlings are the variable. I like to think of them as our drought insurance policy. We run more yearlings in good forage years and fewer in years of lower forage production, ensuring that we keep our grasslands healthy. We are also able to maintain our cowherd in years of lower forage production. Because of this, yearling numbers vary between two hundred and four hundred head. In addition, we grass-finish between one hundred fifty and three hundred head per year.

Managing with Flexibility

We practice what is known as Holistic Planned Grazing (HPG). I want to emphasize several key points on HPG grazing:

- It is goal-orientated.
- It is predicated on stock density, not stocking rate.
- It is not a rigid system or prescription.
- It allows the practitioner to adjust to conditions.
- It is dependent on frequency of moves and frequency of rest.

- It allows for complete plant root system recovery between grazings.
- It allows the practitioner to work with nature, not against it.
- It allows livestock to be used as a soil-building tool.
- Observation by the practitioner is critical to its success.

These, along with the five principles of a healthy soil ecosystem (which I discuss in detail in chapter 7), are key to developing healthy grazing lands.

We typically allow the cattle to consume 30–40 percent of the aboveground biomass. Note that if 50 percent of the aboveground biomass is removed, root growth is not affected. However, if 60 percent is removed, root growth is *cut in half!* This is a very important fact that all graziers need to be aware of. The cattle trample some of the remaining sward, but this varies year to year and livestock class to livestock class. On average, we move the three-hundred-head cowherd once a day during growing season and the yearling herd (between two hundred and four hundred head) anywhere from one to seven times a day. This may sound like a lot of work, but as in any situation, the human mind can make it as easy or complex as it wants. We chose the easy way. The majority of our permanent pastures are 15–40 acres in size. Once a day, a portable fence is set up to divide these pastures even further. These temporary paddocks range in size from one acre to several acres to give us the stock density we desire. We vary stock densities from 50,000 pounds per acre to 700,000 pounds per acre. (To see what this looks like in practice, see plate 12 on page 6.) For those times when we want to move the cattle more than once a day, we use solar-powered automatic gate openers, like Neil Dennis does. We preset a time into each of the gate openers, and the cattle move themselves into the next temporary paddock throughout the day. Talk about stress-free for both the cattle and us!

It is important to note that we do not always move the cattle at this frequency. Whenever we want to take a vacation or some time off, we simply allow them a larger paddock, and we leave them there for a longer time period. This allows us the quality of life we desire.

Most producers allow livestock to graze on cropland only *after* a cash crop has been harvested. I knew that we could advance soil health faster, though, if we made the cover crop our cash crop by harvesting it with livestock *during* the growing season. We do this with different classes of animals at different times of the year, all depending on what our resource concerns are.

I can give you several examples. As I mentioned earlier, I like to grow cereal rye and hairy vetch for their soil-health benefits. They also provide very good forage for livestock early in the spring. Virtually any class or species of livestock will gain weight well while grazing this cover crop. You can also use it to add a thick mat of armor on the soil, which is one of the key principles of soil health. To do so, allow the rye / vetch mix to grow and let the rye mature enough that it starts to produce pollen. Then turn in high stock densities of beef cattle. I prefer to use yearling heifers. The rye is too mature to be high-quality feed, but that is OK. We are not trying to fatten up heifers. They will not relish the rye, but they will eat it and gain a little weight. We allow the heifers to eat only about 25 percent of the aboveground biomass. The remainder is trampled. It usually takes stock densities above 500,000 pounds per acre to get the desired trample effect.

When I use the previously mentioned protocol, I immediately seed another cover crop into that field. I do not use any herbicide when I do this. Usually the allelopathic effect of the rye and the armor it provides are sufficient to inhibit weed growth. On my farm, the two best cover crops for weed prevention in the following crop are cereal rye and sorghum / sudangrass. If there is a nice layer of residue from either of these species on a field, we rarely need to use an herbicide in the subsequent cash crop.

So, what cover crop do I follow the rye and vetch with to graze? That depends, of course, on my resource concern. I need to take into consideration the fact that it will be growing during the hottest time of the year, so the species used must tolerate some heat. Usually, I seed either a cover crop suitable for grazing grass-finishing animals or a cover crop for winter grazing. If I am grazing finishers, I like a mix that is 60–70 percent brown mid-rib sorghum/sudangrass. This is a highly digestible, high-energy warm-season grass. Livestock will select for energy first unless they are deficient in a particular nutrient.

To this I add pearl millet along with cowpeas, mung beans, or soybeans as legumes, a forage brassica such as kale, and at least one flowering species such as buckwheat. I want seven or eight species at minimum, if possible, to take advantage of the synergies in a diverse planting.

We found these annual warm-season cover crop mixes to be an almost ideal ration for the final stages of grass-finishing beef animals. We allow the sorghum/sudangrass mix to reach a height of at least three feet and then graze at densities of around 100,000 to 200,000 pounds of live weight per acre. That is not very high density, and it allows the animals to be more selective. We want them to put on weight.

We usually move them only once or twice each day, and that is more a labor issue than anything else. We move the animals in the afternoon. Why? When will the plant have the highest energy content? In the afternoon, of course, when it is intercepting the most sunlight. Animals select for energy. The animals will strip the leaves off the stalks of the sorghum/sudangrass mix and then take a few bites off the legumes and brassicas. It is not uncommon for us to average 3–4 pounds per day gain on the finishers, which leads to a large amount of intramuscular fat—good fat: high in omega 3, conjugated linoleic acid, and all of the other nutrients that make grass-finished beef so desirable.

Due to the fact that the finishers consume mainly leaves from each plant, the plants will continue to grow as long as it is prior to a frost. As they grow, more carbon is pumped into the soil. Remember, carbon is the key!

What about turnips and radishes? I plant daikon radish to address compaction concerns and to scavenge nitrogen, but not for grazing. They just do not offer much. By the way, if you plant daikon radish before the longest day of the year, the young plants will quickly bolt and go to seed. If you plant them as the day length is declining, they will, with adequate conditions, grow the large tubers they are known for. Turnips are slightly better for grazing than radishes are, but not nearly as good as the forage brassicas— kale or collards.

As you'll recall from the beginning of this chapter, I do not use any fertilizer on any of my crops. My soils are healthy enough to cycle the nutrients the crops need. If your soils are used to synthetics, though, you may need to fertilize your cover crops. I strongly encourage you to cut back on the rate, though. Start healing your soils!

Many people ask whether my animals suffer any problems due to nitrates, bloat, and prussic acid. I can honestly say that we have never lost or treated an animal due to any of these. I can't guarantee that your animals will not have problems, though. I feel our excellent soil health and cropping practices are what helps prevent problems. Bloat is not an issue because of the diversity of the crop mix. Nitrates are not an issue because we have not used synthetic fertilizer for a long time.

People often ask me for a precise "recipe" for a cover crop mix. I purposely do not offer such recipes because what works on my operation may not work on yours. I can share the principles I follow, but you must experiment and find out which species work well in your soils and environment. (I cover the principles and practices of cover cropping that I follow in more detail in chapter 8.)

Since we incorporated Holistic Planned Grazing, the flexibility and options that we have in relation to when and where to move the livestock has greatly increased. We rely on this flexibility for fly and parasite control, which is critical since we no longer use insecticides. They aren't necessary, because we break the fly cycle by moving the cattle away from their manure, which, of course, is where flies lay their eggs. We have also seen a large increase in dung beetle and other insect predator populations. It took two years after we stopped using the insecticides before we saw a dung beetle. Today, Paul has documented seventeen species on our land! Other types of wildlife, such as cowbirds, tree swallows, dragonflies, and a myriad of other predators, keep pests in check. We also run Katahdin/Dorper hair sheep, which act as a dead host for parasites specific to cattle (more about the sheep in *Every Ranch Needs Some Sheep* on page 87). Nature has this figured out; we just have to be smart enough to take advantage of it!

Another of the many benefits of HPG is that it allows us to use the cattle to control noxious weeds. For example, we rented grazing land that had been in the USDA's Conservation Reserve Program for more than twenty years and was primarily composed of smooth bromegrass, a small amount of alfalfa, and a large extent of noxious weeds. We weren't worried at all. At higher stock densities, cattle behavior changes, and they will readily consume less desirable species such as Canada thistle. We have been able to greatly reduce the infestations of noxious and exotic weeds while at the same time increase the diversity and health of other grasses and forbs. Along with Canada thistle, our cattle tolerate grazing on absinth wormwood and even leafy spurge.

Putting It All Together

So, what are the results of my management? Am I truly regenerating soils? Where has all of this taken our ranch?

To demonstrate my belief in the power of regenerative agriculture, I decided to try and quantify the differences that regenerative management has made on our ranch. And fortunately, one of the benefits of traveling around the world to give presentations on regenerative agriculture is that I get to meet a lot of scientists and researchers. Through these connections, we put together a demonstration to study the effects of our management practices and compare them to other styles of farm management.

Four farms were selected, of which ours was one. These four farms had the same soil types and were in close proximity (in order to mitigate weather variables). Here is a brief explanation of the management of each farm.

Farm 1: Diverse Cash Grain Operation

This is a diverse cash grain operation that relies on tillage to prepare the soil for seeding and weed control. Tillage is also used during the growing season to cultivate row crops. Spring wheat, barley, oats, flax, soybeans, dry edible beans, and sunflowers are all grown. Cover crops such as sweetclover are grown and then plowed under in order to provide nutrients to the following cash crop. Natural sources of soil amendments are used. No synthetic fertilizers, pesticides, herbicides, or fungicides are used. This producer does not have any livestock.

Farm 2: Minimum Tillage Operation

This producer uses minimum tillage to grow primarily flax and spring wheat. Once in a great while sunflowers are grown. An air seeder with points is used to apply anhydrous ammonia at seeding. No other form of synthetic fertilizer is applied. Herbicides, pesticides, and fungicides are used when needed. No livestock are owned or integrated.

Farm 3: Medium-Diversity, No-Till Operation

This operation has practiced no-till for many years and has medium diversity in their crop rotation, which includes corn, sunflowers, malting barley, soybeans, and spring wheat. Large quantities of synthetic fertilizers, herbicides, pesticides, and fungicides are used to maximize yields. No livestock are owned or integrated.

Farm 4: Brown's Ranch

The fourth operation is mine. The operation is no-till, with high cash crop and cover crop diversity, and no synthetic fertilizer, fungicides, or pesticides are applied. Livestock are integrated onto the cropland.

Soil samples and water infiltration tests were taken on the same day on each farm. Dr. Rick Haney tested the samples at the USDA Agricultural Research Service Grassland Soil and Water Research Laboratory in Temple, Texas. The results of those tests are presented in table 4.1. Water extractible organic carbon (WEOC) is the food that soil biology eats. Think of it this way: Organic matter is the house that biology lives in, and WEOC is the refrigerator in that house.

As you look at the test results, what stands out? The first thing you'll probably notice is that Farm 4 (Brown's Ranch) has higher soil nutrient levels and a more favorable organic matter level, carbon content, and water infiltration rate than the other farms. But it's also significant to notice how little difference in values there are between the other three operations.

Several key points are supported by the results of this demonstration:

- Tillage is detrimental to all aspects of soil health.
- Low diversity is detrimental to soil health.
- High use of synthetics is detrimental to soil health.
- Livestock integration has a positive impact on soil health.

These data show how crucial it is to manage our farms and ranches as ecosystems. This is absolutely key to healing our families, our farms, our communities, and our planet!

Table 4.1. Soil Test Results for Comparative Farm Study

Operation	N (pound)	P (pound)	K (pound)	WEOC (ppm)	OM (percent)	INFIL (inches per hour)
Farm 1	2	156	95	233	1.7	0.5
Farm 2	27	244	136	239	1.7	0.7
Farm 3	37	217	199	262	1.5	0.45
Farm 4	281	1006	1749	1095	6.9	30.0+

N = Nitrogen; P = Phosphorus; K = Potassium; WEOC = Water Extractable Organic Carbon; OM = Organic Matter; INFIL = Infiltration Rate

I learned the importance of this from the late Jerry Brunetti. In his landmark book, *The Farm as Ecosystem*, Jerry eloquently explained the importance of managing one's farm or ranch as an ecosystem. I am forever grateful to him for what he taught me, especially about how to observe nature. Let nature teach you, through plants, animals, and the soil.

Five

The Next Generation, Building for the Future

Our son, Paul, inherited my love of ranching. Shelly and I had known for years that he wanted to take over the ranch. As a matter of fact, Paul begged us to let him ranch full time right out of high school. We refused because we wanted him to experience some time on his own, even if only for a little while. Reluctantly, he agreed and went off to college. I will never forget Paul's late-night phone calls. He would complain, "Dad, they are teaching the wrong principles! All they want to teach is everything we have quit doing!" I would listen sympathetically and then laugh once he hung up. That perspective was exactly why we wanted him to go to college! Once Paul graduated, he returned home to the ranch to become a full partner in the operation.

We were fortunate that, several years earlier, we had been able to purchase the remaining land owned by Shelly's aunt Alice and uncle Dan. This tract included a small, old farmhouse, which we renovated with new windows, doors, roof, carpeting, heating system, and siding. It was a perfect bachelor pad for a twenty-two-year-old male. The farmhouse is located five miles from our farmstead: out of sight, out of mind.

Several of my farming and ranching friends told me we were making a big mistake by letting Paul come back to the ranch immediately upon graduating from college. They insisted that it would be better for both him and us if he worked elsewhere for a period of time. They said young adults needed to try another occupation in order to be sure they wanted to farm or ranch before returning home. Shelly and I thought about what they said, but it is our belief that if a young person was brought up in an environment that encouraged independent thinking and if that child had been taught how to handle responsibilities, then he should be allowed to make his own career decisions. We do not regret allowing Paul to come back to the operation upon graduating.

I find it very interesting that none of my friends, the same ones who thought we should "force" Paul to take an off-farm job, have a son or daughter who has returned to their operation! It is also interesting that very few of Paul's classmates who grew up on a farm are back home now on that farm. It is unfortunate that Paul is an anomaly. How sad! I believe this is a result of today's agricultural production model, as I explain in more detail in chapter 10.

At a conference I once attended, a speaker posed this question: "How many of you have a son, daughter, or relative taking over or planning to take over your operation?" Of the nearly two hundred people in the audience, guess how many raised their hands? Two: me and one other person. I was stunned! I had expected at least thirty and was hopeful for many more. I find it hard to believe that of the over 120 operations represented that day, only two had daughters, sons, nieces, or nephews that wanted to make a living on the farm. Please don't misunderstand me. I realize that many children do not wish to make production agriculture their career. That is perfectly fine. Everyone should follow their dreams. However, why did only two operations have the climate that was conducive to bringing youth into the business? I thought a lot about this after that conference, so I began asking as many operators and youth as I could this question. What I found were two common reasons: First, there was not enough income on the farm to support two families. Second, it was just too expensive to enter production agriculture. I believe that the solutions to both problems are closely related.

Our Plan for the Future

I have described how my in-laws surprised Shelly and I when, eight years after asking us to work with them and take over the operation, they suddenly decided to split the farm three ways and sell each of their daughters a third. That decision essentially meant that Shelly and I ended up buying not only her third, but her sisters' thirds as well. We had to pay them rent, which covered their mortgage payments. I appreciate the fact that Shelly's sisters allowed us to rent their land—it's a valuable part of our operation. But this scenario weighed heavily in the decision about how we would transition *our* ranch on to *our* children.

Over the years, I had seen too many situations where one sibling returned to the farm while the others pursued other careers, but the parents had not set up an estate plan. Time flew by and suddenly one or both parents passed away and the farm got divided up among all the children. That left the child who spent their adult life on the farm in the difficult position of having to buy out his or her siblings or, worse yet, being forced off the operation. We decided not to leave Paul in that situation. Before he was even out of college, let alone back on the operation, we wanted him to know what our estate plan was.

We sat down with Paul and our daughter, Kelly, and talked things through with them. Kelly was not interested in returning to the ranch. If she had been, we would have made it work so both could have joined us. We put all the land in an income-bearing living trust. The trust owns the land, and Shelly and I are entitled to the income from the trust until we die or until we terminate it. Paul will receive the deeds to the land upon our deaths or upon us terminating the trust. He is also the executor of the trust. In the meantime, Paul is paid a wage from us and he can rest assured that the ranch will be his.

For those wondering about the arrangement with Kelly, upon our passing she will receive all of our personal assets and investments along with our life insurance policies. Is it equal monetarily to what Paul will receive? No. Is it fair? Absolutely. Paul has worked hard with

Planning for the Long, Long Haul

I believe that a farm or ranch is not truly sustainable until it is transferred to the next generation, family member or not. One of the issues in agriculture today is that land has become very fragmented. Children leave the farm, parents pass away, and the children sell off their inheritance, one parcel at a time. It is becoming very difficult to put together a contiguous operation of any size. This is one reason why we have a two-hundred-year plan for this ranch.

Here are the components of our plan:

- Placed the land in a trust to ensure it remains as one unit and is transferred to the next generation.
- Formed an LLC to market our products.
- Invested in a slaughter facility to ensure we are able to harvest our animals.
- Invested in the Bisman Food Co-op, which is an outlet for our products.

us for years already, and he will spend his adult life adding equity to the ranch. He deserves to be compensated for that.

I know some may be asking "What about your retirement? What are you going to live off?" My answer to that is simple: If I can't make enough money while ranching to set aside for retirement, I should not be ranching! I think it is ridiculous that the next generation has to buy the business—one that they helped build. If we run our farms and ranches like businesses, this should not have to happen.

I think it is also important to note that this type of an arrangement assures, as best one can, that if a health crisis develops, a hospital or

- Aggressively work to restore ecosystem function, including the water and mineral cycles. This will help to ensure productivity and profitability.
- Monitor ecosystem function in depth with the support of LandStream.
- Develop fruit and nut orchards to provide income for future generations.
- Continue to diversify revenue streams to ensure a viable business model.
- Seed cropland acres that are close to housing developments to diverse perennials so the land can be grazed, because proximity to urban setting is not conducive to cropping.
- Continue to spin off enterprises to interns in order to both get more people involved in production agriculture and to provide more products for our marketing business.

nursing home cannot force a sale of the land. With the way health care costs are headed, this is an important consideration.

Diversifying Our Livestock

Even while Paul was still in college, he was urging me to consider changes in our operation. One of those late-night phone calls from Paul ended up changing our ranch in a big way. Paul said, "Dad, you are always preaching diversity, but the only livestock we have is beef cattle. I want to get chickens and sheep and maybe even pigs." Whoa! Chickens, sheep, pigs? I had not thought about those possibilities, but it didn't take me more than a few seconds to reply. "Sounds good to me!" After all, my son was right: Diversity is best all around, not just in crops. We immediately started making plans to purchase laying hens when Paul returned to the ranch after he graduated.

Upon his return home, Paul found a lady who wanted to sell her flock of 150 Leghorns. A deal was struck, and Paul was officially in the egg business. The first thing he did was to figure out how to house the birds. He came up with the idea of remodeling an old six foot by sixteen foot stock trailer as a portable chicken coop that could be moved across pastures. We purchased a trailer, and the retrofitting began. We ripped out the old wooden floor and put expanded metal mesh in its place. This allowed the chicken droppings to fall through the mesh and fertilize the land. Then we installed four rows of roosts for the birds on each of the trailer's side walls. We mounted a fifty-five-gallon drum in the nose of the trailer, to which a gravity-flow waterer was attached. To fill the drum, we simply have to pull up to one of the risers in our shallow water pipeline, attach a garden hose, and fill. Last but not least, we mounted a series of nest boxes inside the rear door, which made collecting eggs a breeze.

A couple of additions to the trailer have made the operation more efficient. One is a dolly, which is nothing more than a two-wheeled axle with a two-inch metal ball welded on the top center of the axle. We can rest the trailer hitch on that metal ball. We also attached a metal tongue with a loop in it to the axle, and we can drop that loop over the ball hitch on our ATV. (See plate 31 on page 15.) This makes moving the trailer a simple process; we don't have to monkey around with jacking the trailer up and down. The other addition was a photosensitive door that automatically opens in the morning when the sun comes up and closes in the evening when the sun goes down. In addition to keeping predators out of the trailer at night, it saves us the time and labor of having to lock the chickens in every evening and let them out again in the morning. And what did Paul call this new hen house contraption? Eggmobile, of course!

But what to do with all of those eggs? Paul notified his aunts, uncles, friends, and distant relatives that he had eggs for sale. News spread fast, and soon the demand outweighed supply. Paul called a local vegetable CSA and asked whether he could coordinate with them to set up a weekly delivery point in Bismarck, and they agreed. Paul was so excited each week when he packed up eggs to deliver to Bismarck, and I didn't understand why—until the week I insisted on going with him. When he pulled up to the delivery spot, within minutes more than a dozen vehicles pulled up and young, single ladies (most were single, anyway) got out and surrounded him! (Did I tell you Paul is single?) Well played, son!

I need to explain one sad point about Paul's egg business. You see, what Paul was doing was actually illegal in North Dakota at that time. It was illegal to sell eggs that had not been inspected and certified by the state of North Dakota. This was a case of regulations gone overboard: Individuals should have the right to sell a healthy, nutrient-dense food without government interference! Unfortunately, this is just one of many examples of the screwed-up food system in this country. It's going to take the hard work and advocacy of many farmers and consumers like us—and like you—to bring about change for the better.

Ramping Up Egg Production

Paul's laying hen enterprise has expanded from 150 hens to over 1,100, housed in a fleet of seven eggmobiles. No more washing eggs by hand. We had to upgrade to a commercial-sized state-inspected egg-washing machine that makes the process easy. Eggs from factory-style farms sell for approximately 60 cents a dozen. Paul charges $4.50 per dozen, and people happily pay that price because they want nutrient-dense food. Eggs have been a great entry-level product to draw customers' attention; once they see and taste the quality of our free-range eggs, customers get excited about purchasing other products from us.

How do we manage all of those laying hens? They are the ranch's sanitation crew. During the spring, summer, and fall we run them on pasture, following about three days behind the grass-finished beef herd. This allows just enough time for fly larvae to develop in the manure pats. Dinner is served to the laying hens! Chickens are wonderful creatures; as omnivores they will eat almost anything: clover, grass, flies and fly larvae, grasshoppers, mice, even snakes! It doesn't matter to them. We do feed them some grain screenings (the cracked and broken kernels from our grain crop along with any weed seeds that may have been combined with the grain). A very important way to generate real profit on a farm is by taking the waste stream of one enterprise to fuel the profit in another. Using the grain screenings, which would normally just be waste, to feed the chickens is a good example of this principle. The only purchased supplement we provide to the chickens is oyster shells, which provide extra calcium to ensure strong eggshells.

Due to our severe winters, we cannot house the hens in the eggmobiles during the cold winter months, so in 2015 we built a thirty-six-foot wide by seventy-two-foot long hoop house. Before we move the hens in there for the winter, we add a deep layer of woodchips to balance the carbon:nitrogen ratio of their wonderful droppings. The following spring the house is then cleaned out, and we allow that excellent fertilizer to compost for a year. This compost is then spread on our permanent gardens. Solar energy along with body heat from 1,100 chickens keeps them warm even during the coldest of winter days. The hens are fed a wintertime ration of grain screenings and meat scraps. Yum! They have a good life. As a matter of fact, most industrially raised chickens have a life span of about one year. We have some hens that are seven years old!

A few years ago, a film crew from *National Geographic* spent three days filming on our ranch. The producer asked me if there really was a difference between eggs from truly free-range hens like ours (they could walk to town if they wanted to) and store-bought eggs. I took this as a teaching opportunity and told him to go buy a dozen eggs from any supplier he wanted. The next day he showed up with a dozen organic, cage-free eggs. I then took him out to the eggmobiles and let him pick the eggs of his choice. We took them back to the house, where he cracked open one of the cage-free eggs into a cast iron skillet. It was a typical egg—pale, yellow yolk and watery egg white. He took one of our eggs and cracked it open alongside the other egg. The look on his face was priceless! The yolk from our egg was bright orange, and the white was firm. Excitedly, he cracked open another cage-free egg, same pale color. Another one of ours and another bright orange yolk. He was sold. Lesson taught and learned.

Every Ranch Needs Some Sheep

After starting with chickens, Paul wanted to add sheep. In anticipation, he fenced 320 acres of pasture with three strands of high-tensile electric fence. We thought this would be sufficient to manage the sheep properly. To say that we had no experience with sheep would be a big understatement; neither one of us had ever worked with sheep before. After some research, Paul decided that Katahdin hair sheep, which shed their winter coats in the spring and are known for good meat quality, would be the best fit for our operation. Meat quality was our top priority when making this decision. Having hair was a plus as far as we were concerned, because it meant we didn't have to worry about shearing the animals, but we also lost the opportunity to capitalize on wool as an income stream.

Paul found a rancher a couple of hundred miles north of us who was willing to sell twenty ewe lambs and a ram lamb. Paul chose to buy from this breeder because he was not treating his sheep with wormers or vaccines and had a similar environment to our own. Therefore, the sheep would fit well with how we planned to manage them and would already be adapted to our climate. The ram was turned in with the ewes in December so that they would lamb in May, in sync with nature. I often tell people that we have to learn every lesson the hard way, and we sure learned one when we lambed those ewes out on pasture. Coyotes! They had their fair share of lamb for dinner. A livestock guard dog was added, and that problem was solved.

Early on, I liked to tell people that our sheep were on a planned grazing system—wherever they planned on going, they went there! We quickly learned that the three strands of high-tensile wire was not going to do the trick, so we added a couple of additional wires. Over time, we have continued to learn more about fencing for sheep and about how to handle them.

We added Dorper rams to the flock to add more weight and hybrid vigor. The cross of Khatadin and Dorper has worked well for us, typically yielding a 70-pound carcass at a year of age. In terms of feeding and health care, we manage the sheep just like we do our cowherd—no vaccines, wormers, or grain. They lamb on their own and breed back, otherwise they are sold and gone, no exceptions. They have adapted well, and we have found that the saying "sheep just look for a reason to die" is not true at all!

One of the beautiful things about running sheep is that they tend to eat different plant species than our cattle. This allows us to run sheep without having to reduce the size of our cattle herd. A ranch our size could easily run several thousand ewes along with several hundred cows. However, we have chosen to grow our flock as our market for grass-finished lamb grows.

Hog Heaven

Shelly has a magnet on our refrigerator that reads, "Either you enjoy bacon, or you are wrong!" I, for one, certainly do not disagree with that statement! As we continued growing our meat business, customers kept asking us to supply pastured pork. After doing some research on which breeds of hogs do well on pasture, we selected Tamworths for their foraging ability, along with Berkshires for meat quality. In 2014, we purchased four Berkshire sows, two Tamworth gilts, and a Tamworth boar, and the school of pork-producing hard knocks began!

Large-scale hog farming is another example of what is so wrong with today's production model. The hogs are raised in confinement, and they quickly lose their natural instincts, such as how to take care of their young. Because our initial stock came from the genetic base of animals raised in confinement, we had to go through the same process of repeated culling that we did with the cattle and sheep in order to develop animals suited to our conditions. I think we would have been better off to go to Texas, catch some wild hogs, and bring them back to our ranch as seed stock. Our hogs' foraging ability, maternal instincts, and litter size has improved over time, partly due to the fact that we have added some Gloucestershire Old Spots.

We farrow around twenty sows from April through October. It makes no sense to farrow in the winter in North Dakota, especially considering the fact that our hogs are not confined. In the winter, all we do is put out a few large, round straw bales with the pigs. They burrow right in and stay warm!

During the spring, summer, and fall we run the growing and finishing hogs on perennial pastures. We prefer pastures that have a good legume component. The hogs are supplemented with homegrown ground corn, barley, peas, and oats, plus grain screenings. We have found that grazing hogs on pastures where we bale-grazed cattle during the winter is an excellent way to spread out the residue that remains. The hogs love to root through that residue, eating fungi and insects that have made a home in the piles of carbon (uneaten hay). Why harrow those piles of uneaten hay and manure when the hogs can do it for us? We also use the hogs to "renovate" old tree shelterbelts. They root through all of the old, decaying wood and stir up the area. That rooting stimulates germination of grasses, forbs, and legumes, creating a much healthier ecosystem.

People often ask me whether we graze our hogs on cover crops. We have tried that, and the hogs love the crops, but they can be very destructive to the fields. Even if we had enough labor available to move them often enough (at least once every two days), their rooting behavior might still result in rutted areas that would be difficult to seed into the following year. For this reason, we chose not to follow this practice.

The best thing about raising hogs, other than the bacon and pork chops, is the economic return. Our hogs finish in seven months and provide superior meat quality. Per dollar invested, on our operation, hogs are second only to honey.

Shelly's refrigerator magnet is correct!

Teaching the Younger Generation

Our partnership with Paul is a fundamental part of the success of our ranch, and my family and I believe it is important for us to help the next generation get started in farming and ranching. It always frustrates and disappoints me to see farmers and ranchers who do not want to help the next generation. You can see this situation at nearly every farm auction. A beginning farmer will be bidding on an item, only to be outbid by a well-established farmer. The same thing happens at land auctions. Usually the well-established farmer already has plenty of land and equipment. What a shame.

One way we assist the next generation in getting started is the Brown's Ranch internship program. For the past twenty-five years we have had young people working for and with us during the summer. Many years ago this evolved into an internship program. Each year, we accept applications, conduct interviews, and then award those internships to young people who we feel have the passion and desire to make regenerative agriculture their career. Most years we have a large number of applicants to consider.

We normally accept applications from December 1 to January 15. Shelly, Paul, and I each review the applications on our own and rank each one as a 1, 2, or 3. If all three of us rank an applicant a 1, then that applicant will be interviewed. We like to interview about three times as many applicants as positions we have open. During the interview I like to ask questions that make the applicants think. My favorite is: Would you rather be good and on time, or perfect and late? Those who know me know that, in my world, if you are on time, you are late. I do not tolerate people who are not on time! Interns are selected based on enthusiasm, drive, and desire. We can teach principles and tasks, but we cannot teach those human traits.

Internships normally run from mid-April to mid-October. We provide a small wage and living quarters. Our interns are allowed to eat anything we produce on the ranch, although I do frown on them eating tenderloins every day!

Because of our belief in helping the next generation get started, we also offer an opportunity to our interns that many other internship programs do not. After interning with us, if an individual has shown us that they have the drive, determination, and desire to get into production agriculture, we will spin off part of an enterprise to them. In other words, we sell the intern part of an enterprise of their choice, whether chickens, beef, hogs, sheep, or vegetables. We finance them, provide the land base needed for the enterprise at a reasonable rate, and we buy the product or animals from them, at a predetermined price, to sell through our marketing business. As long as the intern takes care of the enterprise, he or she makes a nice profit. They can continue to work for us as an apprentice while growing their enterprise and accumulating cash. This teaches them business principles and gives them a solid foundation when they do move on.

I try to teach our interns that, when starting out, it's wise to make an operation portable. By *portable* I mean do not buy land or invest in infrastructure. Start with enterprises that can be easily moved to a new location. Grow your operation from profits earned. Build your clientele. Grow as your client base grows. This provides the ability to upgrade to a more desirable situation or location as the opportunity arises. Joel Salatin writes about this in *Fields of Farmers*, a book I recommend to anyone interested in getting started in production agriculture. It is a good read.

I also try hard to educate our interns about the importance of "knowing their Why." The concept of knowing your purpose and sticking to it at all times is crucial in direct marketing, or any other part of business, for that matter. Simon Sinek provides lots of details about this in his great book *Start with Why*. At Brown's Ranch, our "why" is to produce nutrient-dense food while regenerating our ecosystems. Before we make any decision on our ranch we first ask ourselves, "Does this decision hold true to our Why?" Asking that question makes decision making much easier.

Working with interns can be one of the most enjoyable parts of what I do, but it can also be one of the most frustrating. The interns at Brown's Ranch over the years have amazed me with their ingenuity, their recklessness, and their general ability to mix things up in ways I could never have imagined. And this book would not be complete without sharing a few of my favorite intern stories.

One morning many years ago, an intern called me asking, "How do you turn the water off in the pasture?" He proceeded to tell me that a yearling must have caught its hoof against a riser in the waterline and snapped it off, because water was shooting everywhere. The answer would have been complicated, so I told the intern that I would take care of it, and he shouldn't worry about it. Later, when I went out to fix the problem, I found that he had attempted to stop the flow of water in an unusual way. He had stuffed a small zucchini into the broken pipe! I busted a gut laughing! I did give him credit for trying!

One nice summer day, I took an intern to a 150-acre hayfield and showed her how to stack the round bales so that a hay mover could later move the stacks to a spot where we would use them for bale grazing during the winter. The tractor she was using for this task was a John Deere 7220, which has a cab and air conditioning. I thought this would be a pretty easy gig for her, and it didn't take long before she felt comfortable with the task, so I left. Later that day, she called me, crying so hard she could hardly tell me what happened. Between sobs, she explained that she tore the door off the tractor! What?? This was an open field, there wasn't a tree or post in it. And since the tractor has air conditioning, why would she have a door open while driving? I'll never know why she had the door on the tractor open, but she did, and she drove alongside a bale and proceeded to rip the door off. That was a tough one to laugh off.

One week into a new internship season, an intern came running into the shop where Paul and I were working. "They rolled the pickup!" she exclaimed. "Are they OK?" I asked. "One is bleeding, but they are both conscious," she replied. A trip to the emergency room ended with relief that neither intern was seriously hurt. That was very fortunate; I believe that neither of them was wearing a seat belt. The pickup was a total loss. Note to self: When you accept interns from urban areas, take time to teach them how to drive on gravel roads. As Shelly says, "You have to explain *everything* to them."

Sometimes, even thorough explanation doesn't help. One summer I had planted a 1-acre patch of sweet corn in the middle of 60 acres of field corn. I showed an intern the patch of sweet corn; the boundaries of the patch were clearly marked with bright orange flags. She and I start weeding the patch by hand (I refuse to use herbicide on any crop that will be consumed for food, by either humans or animals). After a while, we headed home for the day. The next day I sent her over to finish the job. She did not return until suppertime. *What a slow worker*, I thought. That evening she posted a photo on Facebook, saying that she was working awfully hard and it was going to take her a long time to weed the corn. She was standing in the middle of the field corn! She had forgotten that the job was to weed the patch of sweet corn! Yikes! You just can't make this stuff up.

Last story: I had helped an intern fill the seed drill with cover crop seed and left him on his own to finish seeding a field. I told him to call me when he was done with that field so I could meet him at the next one. Later in the day he called saying he was done and was on his way to the next field. I loaded some seed into the back of my pickup and headed to that field. I got there, but he was nowhere in sight. I waited, and waited. The field was only two miles from the field he had just seeded. I waited some more. Finally, I saw him coming down the road—he had the tractor in field gear! Evidently, he didn't know how to shift to a higher gear.

Upon his arrival I jumped up on the drill so I could add seed. Much to my dismay, the drill was empty! "You have an empty drill," I emphatically stated. "Oh, don't worry," he said, "there was plenty of seed in it when I started." Yikes! I knew what that meant, and sure enough, when the plants started to emerge, it was apparent that he had run out of seed with 15 acres to go. Lesson learned, hopefully. I often think interns are God's way of teaching me patience.

Six

Nourished by Nature

Experience has taught me—and many other farmers will agree—there is no money to be made producing commodities without accepting taxpayer subsidies. Producers typically earn only fourteen cents out of every dollar the consumer spends on food products. Why would I be satisfied with only fourteen cents for my products when there is another eighty-six cents out there? Yes, I am a capitalist! And once Paul joined the business, we realized that in order to reach our ranch goals, we had to develop a business model that would help us capture the eighty-six cents that conventional producers are losing out on. That meant direct marketing our farm and ranch products.

The biggest challenge facing most producers who want to get into direct marketing their meat is the processing, and that was true for us, too. When we first started raising grass-finished beef, there were only four slaughter facilities in the entire state of North Dakota that were inspected to allow the retail sale of processed animals. The waiting list to get any livestock processed in these facilities was thirteen months. We realized that we could not operate with those constraints, so in 2012 we joined a group of producers and other investors to form Bowdon Meat Processing (BMP), a cooperative that is inspected by the state department of agriculture to allow for retail sales of meat. It is located in the small town of Bowdon, North Dakota, which is ninety-three miles away from our ranch.

The location was chosen because the town previously had a meat-processing plant that had been forced to close after the owner suddenly died and the building could no longer meet regulations. After $1.3 million was raised through investments, grants, and loans, the building was demolished and a new one was constructed. BMP opened its doors in April of 2014. The plant was located in Bowdon for a number of reasons. The first was that the community wanted it there and was willing to work with the co-op to get the necessary zoning permits, something that would not be easy to do in a larger community. Another was that many of the people in the town were willing to invest in the co-op if the facility was located there.

The BMP plant slaughters and processes buffalo, beef, lamb, hogs, and goats under state inspection. It is a relatively small facility, with a weekly capacity of about twenty-five head. Like any small plant, it has to charge more per head as compared to large plants due to economies of scale. To make up for our higher processing costs, we have to charge more for our product. We do this by selling nutrition, not commodities.

We have a standing appointment to have our livestock processed at BMP every two weeks. This allows us to choose on an ongoing basis which species of animals we need to have processed to meet our inventory needs. We have discovered that it's imperative to maintain an accurate record of our meat inventory to make sure that we do not run out of any products. We transport the animals to Bowdon, and on the same run we bring back home the meat that was processed from the animals delivered two weeks earlier. It works out well for us.

While BMP was being constructed in 2013, we set up a separate business entity, Brown's Marketing LLC, with the guidance of our lawyer. The purpose of this company is to retail or wholesale any products produced on our ranch. Brown's Marketing LLC purchases the animals or produce from Brown's Ranch for a predetermined price, has them processed, and then retails those products. This business structure is important because it takes the liability risk away from the ranching entity. I refuse to risk losing the ranch through a lawsuit if, for example, a customer gets sick because someone did not follow safe meat-handling practices, such as failing to meet safe handling protocol when preparing one of our products.

We set up this LLC with Paul as 60 percent owner and Shelly and me as 40 percent owners. Paul is the president and makes all of the business decisions. We chose this structure purposefully to teach him both the financial and marketing aspects of a business. I have seen too many farm families where one or both of the parents make all of the decisions and take care of the financials. They may have a son or daughter who is fifty years old and has no idea how to run a business. Shame on those parents!

Once the LLC was formed, we focused on upscaling our marketing and growing the business. We decided we needed a trademark for the LLC that would be appealing to the consumer and convey what we are about. After much consideration, we settled on *Nourished by Nature*. It was simple and yet it encompassed our goals. We applied for and were granted that trademark so that our products could be merchandized under the Nourished by Nature label.

Infrastructure for Retail

With processing dates scheduled at BMP in the spring of 2014, we needed a home for the finished products once we picked them up from the processor. We settled on an old building on the farm that Shelly's parents had built in the late 1950s to house laying hens. (That endeavor only lasted a few years before regulations and depredation from fox and coyotes caused its demise.) Over the decades, the building had been used for storage. First, we gutted it. Then we rewired it, added plenty of outlets and lights, and sprayed three inches of foam insulation on the interior walls and ceiling. Then we purchased a few chest freezers and refrigerators and set them up in the building. At that point Paul had an efficient space for housing the inventory and running the direct-marketing business. It took an initial investment of $10,000 to start up this business. Since that time, though, the business has ready cash flow and has not had to borrow a penny. There is a real lesson to be learned here: Grow a business with profits, not with debt!

The next purchase Brown's Marketing made was a concession trailer that could be pulled to farmers markets and to the vegetable CSA and egg delivery site. The trailer was custom-designed to house two large chest freezers for meat and a refrigerator for eggs and produce. It also has a compartment that houses a generator and a countertop for carrying out transactions with customers. (See plate 28 on page 13.) In June 2014, we pulled into our first farmers market in Bismarck with high hopes, not knowing quite what to expect. WOW! The demand for our products was greater than we had anticipated! It didn't take us long to realize that we had filled a niche within our community. Their support was very encouraging and proved to us that people *do* want to buy nutrients, not commodities.

During the summer of 2014, Paul started looking at his options for the winter season. Since farmers markets and vegetable CSAs last only about five months in North Dakota, there is a long "off season." Fortunately, we already had a lead on how we could continue sales during that long season without markets. The year before, we had met Blaine Hitzfield and his father, Lee, when they shared the story of their marketing model at the Grassfed Exchange Conference that was held in Bismarck. We knew it was exactly what we needed to fill our void, and we reached out to Blaine to see if he could help us get started. This contact proved to be crucial for the success of our direct-marketing business.

Blaine is one of seven sons who operate Seven Sons Farm outside of Roanoke, Indiana. Their story is similar to ours: Lee was a conventional producer who decided to change their farm management in the early 2000s. Today, the farm direct-markets their pastured products to over five thousand families throughout the

Midwest via scheduled deliveries. Blaine had mentioned that they had developed a software called GrazeCart for their own business and were working on expanding the platform so that other direct marketers could use it for a monthly fee. Bingo! Paul began working on a new website using GrazeCart software that would allow us to set up scheduled deliveries so that our customers could have access to our products year-round. Our website www.nourishedbynature .us was launched in the fall of 2014.

The software has a user-friendly interface that allows us to add both products and product categories, track inventory, and set up delivery points and shipping options. Developing a website that is easy for customers to access and make purchases through is very important. After all, everyone is online these days! When a new customer first visits our website, they are prompted to enter their information and create an account. From there, they choose a delivery location where they will pick up their products at a designated delivery date and time. After they choose their location, they can fill their cart with the products of their choice and submit their order. Customers are allowed to order anytime up to forty-eight hours ahead of the scheduled delivery time. That two-day window gives us ample time to pack the orders before we hit the road. Once we have processed and packed an order, the customer's credit card is charged, and an email is sent to them as a reminder to pick up their purchase at the appropriate time and place. This simple process allows us to know in advance how much revenue we will make on every delivery run. Once we arrive at a delivery location, the customers arrive, pick up their orders, and head home with nutritious food, all within a half hour.

The Customer Is Always Right

We have continued to update the website and our product line in order to keep up with customer feedback. This feedback is perhaps the most important aspect of the relationship that we build with our customer base. There is absolutely nothing like having direct contact with customers to find out what they are thinking and wanting. It has been quite interesting to take note of the questions that are asked most often. Ninety-five percent of the time, the first question a customer asks is "Where are you from?" They simply want to know where the food is raised or grown and where they can find you. We tell them "just seven miles straight east of Bismarck" or "Have you noticed the smiley-face painted rock outside of Bismarck along I-94? Yep, that is home base for our ranch." Creating an open, inviting ranch is important.

The second question (over 80 percent of the time) is "Do you grow and feed any GMOs?" I must say that we are really surprised at how often this question is asked, especially in a rural state such as North Dakota. You can argue the pros and cons of GMOs all you want, but if your customers do not want products that contain GMOs, why would you grow or feed them? The next three questions most often asked (the order varies) are: "Do you feed any antibiotics?" "Do you use any (added) hormones?" and "How do you treat your animals?" Paul answers this last question by telling folks "If I were an animal, I would want to live on Brown's Ranch." I prefer to answer it this way: "Our animals have one great life and one bad moment!"

Taking questions from our customers is great for us because we can then aim our marketing materials at answering them and appeal to those who find these questions important when choosing to spend money to feed their families. Interestingly, it's rare that anyone asks whether our products are organic. To our knowledge, we have never lost a sale because our operation is not certified organic. By taking the time to explain our "why"—our goals and practices—we satisfy our customers. And they are willing to pay prices equal to or greater than those charged for certified organic products.

You must build trust, be transparent, and have product integrity in order to create a reputable brand. As I mentioned earlier, you must know your "why" and be able to portray that throughout every avenue of your business. When you send a clear message, potential and existing customers will know exactly what your business is and what it stands for. This will create an inherent high standard that your product line and business will be known for, ultimately earning trust.

All three of these principles build off one another. For example, we list all ingredients very clearly on product labels, which creates *transparency* with the customer. If the word "spices" shows up in a seasoning mix, we make sure to define what those spices are exactly. We take our customers' health seriously, and to uphold a high standard we choose not to carry products that contain MSG, artificial nitrates, maltodextrin, high-fructose corn syrup, dextrose, or any other additives. After all, we choose to go above and beyond when raising our livestock, so why would we want to ruin that standard of *product integrity* by adding a bunch of unnecessary additives? When customers see this, they realize that we care and have their best interest in mind, which builds *trust*. It always puzzles me to hear producers tout the benefits of their pasture-raised proteins only to see their products contain the very "junk" that they are trying to differentiate themselves from!

Another great way that we build trust with our customers is through our open-door policy (another manifestation of transparency). Our customers know that they are free to visit any time they wish. We simply ask that they give us a heads up so that we can be there to guide them or to point them toward where the livestock are so that they can walk out and look for themselves. There is nothing more rewarding than watching a family's reactions when they visit the ranch and get a chance to see our "why." And hey, their trip usually ends up with them going home with a few Nourished by Nature products in a box, too. Build trust, and your customers will continue to reward you with their food dollars. This is a unique connection that no commodity market or grocery store chain can build. Bridging the producer–consumer gap is of utmost importance to us.

I don't want to paint any false pictures here; direct marketing takes a lot of work. It is all about delivering a high-quality product with great customer service. There are a few things that differentiate Brown's Ranch from most others who direct-market in our area. The first is that we raise our livestock from birth to finish (except for the chickens). The second is that we grow the grains that we feed our chickens and hogs, as well as, obviously, the forages that our cattle and sheep graze. Therefore, we know exactly what our livestock are eating from day one. As our soil health builds, so does the nutrient density of our plants. We take brix readings of our crops, vegetables, and forages. Brix is a measurement of dissolved solids (usually sugars, in the case of crops) in a liquid and is a good indication of nutrient density (the levels of bionutrients in the plant tissues). Brix has been used in the wine industry for years to determine when grapes have reached their peak for sweetness, making them ready for harvesting. The brix readings of our crops, vegetables, and forages have increased substantially over the past ten years, further solidifying the fact that we are growing nutrient-dense foods, not commodities.

We still have a lot to learn, but we continually seek out new ways to build our customer base. Farmers markets have been a good way for us to get exposure and acquire new customers. In 2016, in addition to our trailer, we purchased a cargo van that holds a couple of chest freezers and a refrigerator. This van has been a great addition because we travel up to two hundred miles to attend farmers markets. We like to use the markets as our preorder delivery points, so that we can kill two birds with one stone during the summer and fall. Since we have expanded our market area, we have expanded our customer base and, as of this writing, serve over 1,200 families throughout North Dakota. To accommodate this increased market, we purchased a walk-in freezer. Remember the old chicken coop that was converted to hold the refrigerator and freezers? It is now full with eleven freezers and three refrigerators—another testament to the fact that people do want healthy, nutrient-rich food.

I'm proud of the success of the Brown's Ranch, but I am never completely satisfied and I continue to seek better ways to do things. For example, we continue to diversify our business enterprises. Because we always have plants flowering during the growing season, there are always a large number of pollinator insects taking advantage of the supply of pollen and nectar. So why not produce honey? We did some research and located an apiary that was willing to work with us. They place their hives on our property and the bees go to work pollinating our crops while producing nutritious honey. The apiary extracts the raw, unfiltered honey from the hives and bottles it in 1-pound, 2-pound, 5-pound, or one-gallon containers, which we provide. The owner of the apiary told us that the hives placed on our property yield 19 percent more honey as compared to the hives placed on other properties. I see this high yield as proof of the diversity and health of our ecosystem. We pay them a fair price, thus helping to support a local business. Then we sell the honey to our customers at a small profit. It is a win-win situation for all involved, including the bees!

Earlier I mentioned that as part of our two-hundred-year plan we recently planted fruit trees: over 1,500 apple, pear, peach, plum, apricot, juneberry, cherry, saskatoon, currant, aronia berry, blackberry, blueberry, and mulberry trees and bushes (see plate 34 on page 16). Fruit trees are not something you normally see in North Dakota. I like to tell people that, "people laugh at me because I'm different, but I laugh at them because they are all the same." These fruit trees will allow us to add value in a variety of ways. Obviously, the sale of fresh fruit is an option because very few others have locally grown fruit to sell, but we can also sell cider, hard cider, jams, jellies, pies, and wine. We have also planted chestnut, hazelnut, filbert, and walnut trees, which, although they can take as long as thirty years to fully produce, will leave future generations with another viable economic enterprise.

We are considering other enterprises, as well. Ducks, turkeys, rabbits, dairy, and the list goes on. A person is only limited by his or her imagination. Too many producers overlook potential income streams. It is not all about money, though. One source of income that we choose not to pursue is commercial hunting. Our cover crops and diverse perennial pastures attract a lot of wildlife. Many people are willing to spend a healthy sum for the opportunity to hunt that wildlife, both with a gun and a camera. We choose not to open our land to traditional hunters; instead, we open our ranch to an organization called Sporting Chance, which gives handicapped individuals the opportunity to hunt. Law enforcement and veterans are extended this offer, too.

Have I achieved my goal of capturing the eighty-six cents of every food dollar? Not yet, but we're on our way. What I know for certain is this: A farm cannot be sustainable, let alone regenerative, unless it is profitable. We need to put the profit back into production agriculture!

PART II
The Big Picture

Seven

The Five Principles of Soil Health

I began this book with the principles of soil health, and they weave through the entire story of Brown's Ranch told in part one. Because of their fundamental importance to regenerative agriculture, I return here to explain in more detail why each principle is so important, what happens when each principle is ignored, and how to apply the principles through your farming, ranching, and gardening practices.

These five principles of soil health were developed by nature, over eons of time. They are the same anyplace in the world where the sun shines and plants grow. Gardeners, farmers, and ranchers around the world are using these principles to grow nutrient-rich, deep topsoil with healthy watersheds. I credit Jon Stika (author of *A Soil Owner's Manual*), Jay Fuhrer, and Ray Archuleta for being the first, to my knowledge, to refer to these as the "five principles of soil health."

It is imperative that all farmers and ranchers understand these principles, for to ignore them will lead us farther and farther down the path of complete degradation of all natural resources—not just soil. Without healthy soil, we cannot have healthy crops or healthy animals or healthy people. We must promote the health and functioning of the ecosystems in which we farm. Like humans, nature can handle occasional stress, but, just like humans, nature cannot function properly in the face of prolonged or acute stress.

Principle One: Limit Disturbance

The first principle is to limit mechanical, chemical, and physical disturbance of the soil. Where in nature do we find mechanical tillage? Nowhere, of course!

Humans have been tilling the soil for thousands of years, and as modern technology has increased our ability to till more acreage faster, harder, and deeper, the damage done becomes ever more serious. Widespread tillage may make certain tasks easier for the operator, but it destroys soil structure and function. In his book, *Dirt: The Erosion of Civilizations*, Dr. David Montgomery notes that the demise of civilizations throughout history has been tied to the degradation of their soil resources. The principal contributor to that degradation was, of course, tillage.

Many producers believe that by tilling they improve soil function. Nothing could be further from the truth. Tillage immediately destroys soil aggregates, significantly decreases water infiltration rates, and accelerates the breakdown of organic material, among other effects. During this intrusive process, oxygen is infused into the soil, which stimulates particular types of opportunistic bacteria that quickly multiply and consume the highly soluble carbon-based biotic glues. These highly complex natural glue substances hold the micro and macro aggregates (composed of sand, silt, and clay particles) together. When the glues are gone, the silt and clay particles fill the voids, which reduces porosity. This reduction results in anaerobic conditions in the soil, altering the type of soil biota, which in turn may lead to an increase in pathogens and loss of nitrogen in the system because of an increase in denitrifying bacteria. Carbon dioxide is released into the atmosphere. As microbes die they release soluble forms of nitrate nitrogen into the soil solution, which stimulates weed growth. Tillage also diminishes complex mycorrhizal fungal networks. The severed hyphal network can no longer deliver complex amino acids and other complex organic/inorganic molecules, thus impacting plants, animals, and humans. Fewer nutrients for the plants also means fewer nutrients for animals and people, as I discuss in chapter 10.

This is the main reason the soils on my ranch saw organic matter levels drop from an estimated over 7 percent pre-European settlement to less than 2 percent at the time Shelly and I purchased the land from her parents. Consider that organic matter (carbon) controls 90 percent of soil functions related to plant growth, and you understand why tillage is so destructive.

I have had the good fortune to visit hundreds of farms and ranches all over the world, and I am always distressed by the degradation I see as a result of tillage. Even while on a visit to the farm that many claim has the best soils in Australia, I was disappointed. Dr. Christine Jones informed me that I was actually standing on subsoil because over a meter of topsoil had been lost from that field over time due to tillage! I have toured many fields in the states of Illinois, Indiana, and Iowa that are touted as some of the most productive in the world, saddened by the realization that they are a mere fraction of the deep, rich soils they once were.

Chronic chemical disturbance is just as devastating. The application of copious amounts of fertilizer and herbicides can destroy soil structure and ecosystem function. For more than one hundred years, the Morrow Plots on the Urbana-Champaign campus of the University of Illinois have been a site for study of continuously cropped corn, soybeans, and hay. In 2009 the researchers who manage those plots released a paper titled "The Browning of the Green Revolution," in which they stated that "logically, the soil should gain nitrogen if fertilizer inputs exceed grain removal." However, at the Morrow Plots, despite application of at least 60 percent more nitrogen than the amount removed in corn grain, over time there was a net decline of 624 pounds to more than 1,600 pounds per acre in total soil nitrogen. This situation does not seem sensible, or even possible.

The authors stated, "despite five decades with approximately double the input of synthetic nitrogen, corn yields are still lower with monoculture cropping than with the two rotations. This disparity is consistent with differences in potentially available soil nitrogen, a key factor in sustaining soil productivity. An inexorable conclusion can be drawn: The prevailing system of agriculture does not provide the means to intensify food and fiber production without degrading the soil resource."

That last sentence hits me hard. "The prevailing system of agriculture does not provide the means to intensify food and fiber production *without degrading the soil resource.*" If chemical agriculture is destroying the soil resource, then how can we justify continuing down this path?

You may be puzzled about why the researchers saw the results they did. The answer lies in the plant–soil microbe relationship, as discussed in chapter 3. If we feed a plant water-soluble synthetic fertilizer, that plant, more or less, becomes lazy. It no longer needs to release as much carbon into the soil to attract soil microorganisms. The net result is a decrease in the numbers of beneficial microorganisms and fungi. Less soil life, in turn, means less aggregation, reduced pore space, and lower water infiltration. Included in this cycle is a significant loss in nitrogen-fixing bacteria such as *Azotobacter*. All of this means a degradation in the function of the soil ecosystem.

Herbicide applications can be just as destructive. Dr. Don Huber, professor emeritus at Purdue University and one of the world's leading authorities on chemical–soil–plant interactions, is a wealth of knowledge on the effects of herbicides on the environment. He is alarmed at both the number and the volume of herbicides being applied and their effect on both the ecosystem and human health. Here is a mind-numbing statistic: In 2017, there was enough glyphosate used in the United States to cover *every harvested acre of cropland* at the standard application rate of three-quarters of a pound per acre. (Worldwide there was enough glyphosate used to spray every acre of harvested cropland with two-thirds of a pound.)

Glyphosate is registered as a chelator, which means it binds to metals. Is it possible that glyphosate is tying up nutrients in the soil that could be used by plants? Glyphosate is also registered as a biocide, meaning it kills biology. Can we also deduce that glyphosate is harming soil life? I am not saying that glyphosate is the only culprit. Applying *any* herbicide, fungicide, or pesticide will have a negative impact on some aspect of the environment. If you are a conventional producer, before you get mad and put down this book, please take a moment to think this through. Everything we do in production agriculture has compounding effects. If we apply an insecticide to kill a particular pest insect, the insecticide is not going to kill only that species, it will kill others as well, including some that are harmless and many that are beneficial. We *must* realize that. Nature can handle an occasional stress; in fact, occasional stress can have a positive effect. However, nature cannot handle *chronic* stress, such as the yearly use of tillage, synthetic fertilizers, pesticides, and fungicides.

Principle Two: Armor the Soil Surface

The second principle is to maintain armor (of plant residues) on the soil surface. Where in a healthy ecosystem do you find bare soil? Your first response might be, "Gabe, there are plenty of places where the soil is bare!" Sadly, yes, but is it healthy soil? If bare soil was normal in nature, then why do weeds grow whenever we till an area? Nature is trying to cover the soil! The truth is there should not be many open expanses of bare soil, because bare soil is a sure sign of a dysfunctional ecosystem. I often hear producers who live in drier environments claim that their land has always had some areas of bare soil. But historical records, including old journals, show us that even the areas we now consider deserts were once covered with vast grasslands. Recently in Oklahoma, a farmer told me that his grandmother came in a covered wagon in the 1800s to settle what is now their property. She said that the prairie grasses were so tall then, a man riding on a horse often could not be seen! What an amazing contrast to the reality of the Oklahoma landscape today.

I learned the hard way about the value of armoring the soil during those years of hail back in the 1990s. The hail knocked down the vegetation, and I saw how all those felled plants protected the bare soil. The following year I noticed that this crop residue inhibited weed growth; kept soil temperatures down during the heat of the summer; reduced evaporation rates; and provided valuable soil organic material, which was cycled by the earthworms that seemed to magically appear. This armor is also home to a myriad of microorganisms.

When a raindrop hits plant cover instead of bare soil, much of its energy is dissipated, thus protecting the soil from water erosion. Drive through any crop-growing region anywhere in the world where there is tillage and you will see soil being carried away by the wind. Wind erosion is almost as prevalent today as it was during the Dust Bowl. While writing this book, I made a trip to central Oklahoma, and authorities there had to close an interstate highway because of low visibility due to blowing soil. This is a travesty! Consider this: One ton of topsoil spread across a 1-acre field would have the same thickness as the sheet of paper these words are printed on. Picture that, and then figure: How many tons of topsoil were lost in the wind that day in Oklahoma?

As you reduce mechanical, chemical, and physical disturbance, allowing your soil biology to improve, a new challenge may emerge: keeping up with the continuing need for new armor. As soil health improves, earthworms and other soil biology will cycle through surface residues more and more rapidly. In the years following the hail and drought on our land, the biology in my soil multiplied rapidly. The soils on Brown's Ranch are now so biologically active that I've seen an inch-thick residue disappear in six weeks! I address this issue by increasing the amount of carbon relative to nitrogen in my crop rotation (I discuss this concept more in *Principle Three: Build Diversity*). In practical terms, I reduce the amount of legumes in my cash crop rotation and in my cover crops.

Another way to promote thick armor is to grow a high-carbon cover crop, allow it to mature to the point where it's starting to pollinate, and then graze it at high stock density. I allow the cattle to consume some of the plant material, say 25 percent of the aboveground biomass, but I make sure that they also knock down a thick layer of litter that will armor the soil. (See plate 11 on page 15.) This is important on the native rangelands, too. At all costs avoid overgrazing, which creates bare ground. If bare spots do show up, use livestock impact to help those bare spots recover (I explained this technique in *The Power of Stock Density* in chapter 2, page 34). Bale grazing is also a good way to armor difficult bare spots on your operation.

The fact that armor buffers the temperature fluctuations of the soil benefits both plants and soil biology. Many producers do not pay enough attention to soil temperatures, but temperature can have a dramatic impact on plant health. Consider the following:

- When soil temperature is 70°F (21°C), 100 percent of soil moisture is available for plant growth.
- At 100°F (38°C), only 15 percent is available for growth, the remaining 85 percent is lost due to evaporation and transpiration.
- At 130°F (54°C), 100 percent of the moisture is lost to evaporation and transpiration.
- At 140°F (60°C), soil bacteria die.

As producers, we make our living from growing plants. It's in our best interest to give our plants the best habitat possible, especially below the soil surface. Keeping the soil well-armored should be one of our top priorities.

Principle Three: Build Diversity

The third principle is to promote diversity on as many fronts as possible. My son, Paul, taught range management at the local community college for five years. Each year he brought his students to one of our pastures and had them collect as many different grasses, forbs, and legumes as they could find. One year the students collected over 140 species! That is the level of diversity we find in a natural (well, as natural as it can be in this day and age) ecosystem. Lewis and Clark found that level of diversity when they explored the Missouri River system in the early 1800s, including diversity of plants, animals, and insects. The rich, deep topsoils that once covered large parts of this planet were all developed over time due to this diversity.

Let's consider the current agricultural production model. I can drive for hundreds of miles throughout the Midwest and not see any crop other than corn or soybeans. In the Southeast I see cotton everywhere. In the Pacific Northwest it is wheat. These monocrops are the opposite of diversity.

Once, after I gave a presentation at a conference in Kansas, a young producer approached me and asked me how he could get his father and grandfather to add diversity to their cropping system. "What is your rotation now?" I asked. "Well, since the late 1920s we have never seeded a crop other than wheat," he explained. Wow! Over ninety years of wheat. "It must not be yielding well," I surmised. "No, it doesn't," he replied. "It averages about eighteen bushels an acre, and we all have to hold down off-farm jobs."

Another time, a young producer in Canada asked me how he could explain to his father the importance of diversifying. "What is your crop rotation?" I asked him. His reply: "Canola, snow, canola."

These may sound like extreme examples, but my observations and conversations with producers indicate that they are much more common than you might imagine. Farmers need to pay more attention to the four crop types: cool-season grasses, cool-season broadleaves, warm-season grasses, and warm-season broadleaves. (I discuss these in more detail in chapter 8). Each of these crop types influences a field ecosystem in a different way. If we examine a healthy pasture, we will find examples of all four of these crop types, in varying proportions depending on location. It stands to reason then, that we should have all four of these crop types in our rotations. The vast majority of producers focus only on the potential profit a particular crop may bring them that year; they do not look at the ecological capital that diversity builds. If a pasture ecosystem in its natural state includes as many as one hundred different species of grasses, legumes, and forbs, how can we possibly expect the system to function well if we reduce plant diversity to only one or two species?

If you want to improve your soils, you must add diversity either by diversifying your crop rotation or by adding cover crops. My good friend David Brandt tells everyone who will listen that the biggest improvement to his soils occurred when he added winter wheat to his corn/soybean rotation. The benefit wasn't just from adding wheat; David also planted a diverse cover crop immediately into the wheat stubble after harvest. The cover crop is what made the biggest difference. Instead of the soil biology feeding on root exudates from only two species (corn and soybeans), it feasted on well over a dozen species. Think of the increased amount of carbon cycled due to all of those living plants.

David's results are similar to those of the Burleigh County Soil Conservation District cover crop demonstration in 2006 (described in chapter 2). In that trial, the six-species blend yielded two to three times as much biomass as the single-species cover crops.

Ecologist Dr. David Tilman at the University of Minnesota has done some great work showing that synergies are compounded once plant diversity reaches seven or eight species. In other words, plant health, function, and biomass improve and increase with diversity. One of many important reasons for this is the fact that a diverse plant population provides a much more diverse diet of root exudates for soil microbes. Given this fact, it is even harder to understand why there is still so much support for our current monoculture production model. Dr. Tilman's work also showed that plant health, function, and biomass improved and increased as more functional groups were added, in other words when grasses, forbs, and legumes were planted together.

As you first start to shift to regenerative agriculture practices, you'll probably find that you need to include more legumes in the rotation. This is related to the effect of carbon:nitrogen ratios. The organic matter portion of soil has a carbon:nitrogen ratio (C:N ratio) of about 12:1—twelve parts carbon to one part nitrogen. Plant residues on the soil surface will have a range of C:N ratio, depending on the type of plants. Crops such as cereal rye and wheat have a C:N of approximately 80:1. Corn has a ratio of 57:1; alfalfa 25:1; hairy vetch 11:1.

Whatever the C:N ratio of the surface residue, soil biology will eventually break it down to a ratio of about 12:1. The ideal C:N ratio for easy decomposition of residue is 24:1. This ratio is optimum for the health of microorganisms. When crop residue is too high in carbon, there is not enough nitrogen to support the microorganisms, and they must find other nitrogen sources in the soil. This ratio is important to keep in mind when selecting crop and cover crop rotations.

Often, a producer who switches to a no-till system finds that the residue is slow to break down, due to a dysfunctional nutrient cycle. As a rule of thumb, the higher the C:N ratio of plant material, the longer it will take to decompose. The lower the C:N ratio, the faster the residue will decompose. The ratio also determines how much nitrogen will be tied up and thus not available for the subsequent crop. High-carbon plants, such as wheat, break down much more slowly than low-carbon plants, such as peas. The answer to slow breakdown is to grow cash and cover crops with a C:N ratio that allows for proper nutrient cycling in the soil so that residues cycle in a timely manner.

For example, I have a friend who, after five years of no-till, called me to complain about the fact that such a thick layer of residue had built up, it was difficult to seed a crop through it. I went to his farm to see for myself. Examining the field, I could tell by the residues what had been planted there the last five years: sunflowers, spring wheat, corn, barley, and winter wheat. All of these are high-carbon crops. My friend did not have a residue problem, he had a carbon:nitrogen ratio problem. The solution to his dilemma was simple: add legumes, which are high-nitrogen crops. He added peas to his rotation and also began to seed legumes and daikon radish following any early-harvested cash crops. The legumes helped to balance the C:N ratio, and the radish roots stored that nitrogen and released it the following spring when they decayed. This nitrogen then accelerated the decomposition of the crop residue. Problem solved.

Diversity is important not only for cropland but also for grazing land. I often think back to my first experiments seeding cropland back to perennials to create pasture. I chose a mix of smooth brome grass, along with intermediate and pubescent wheatgrass. Talk about little diversity! No legume, no forb, it is no wonder that stand did not produce much growth.

Principle Four: Keep Living Roots in the Soil

The fourth principle is to maintain living roots in the soil as long as possible throughout the year. I am always disappointed when producers harvest a grain crop and then leave that land sitting idle without any living roots in the soil until the following year. In October 2017, I drove from my home near Bismarck to Butte, Montana, a distance of over 650 miles. How many green growing fields did I see, once I left my ranch? Only one! Only one farmer along that whole stretch had taken the time to plant something after the season's harvest. It was clear that the other growers had not learned the importance of pumping liquid carbon into the soil to sustain soil biology. Here's an analogy: A farmer would never leave their livestock unfed for months at a time. Why, then, do farmers not think to feed their "underground livestock" through the winter? People often ask me, "What is the one thing you have done that has made the biggest difference to your soil?" The answer is simple: "I grow plants!"

Never, ever pass up the opportunity to convert solar energy into biological energy. As soon as I am done harvesting one crop, be it by combining or grazing, I immediately seed another crop or cover crop. Think of how this ties to the nutrient cycle. If we are not pumping liquid carbon into the soil, we are not feeding soil biology; if we are not feeding soil biology, we are not cycling nutrients. Once you understand these simple principles, you will have new insight into why many producers need to use copious amounts of synthetic fertilizer to grow a crop. Their soil's natural fertility has been starved out.

Another very important reason for having livings root in the soil is to enhance and proliferate mycorrhizal fungi. I have already explained the myriad of benefits fungi provide. Why not take advantage of those benefits?

In many areas of the country, cover crops planted after a cash crop will not put on much top growth because the number of frost free days are limited or moisture is limited. Many years, I seed a cover crop following a cash crop only to see it grow perhaps three inches tall before frost kills it. This is not a failure! Even though aboveground growth isn't much, those little plants have produced plenty of roots underground, and that is what matters.

If moisture is an issue in your area, growing plants is even more important because the only way to increase the water-holding capacity in the soil is by increasing organic matter. Approximately two-thirds of any increase in organic matter is due to roots. It is critically important to have as many roots in the soil as long as possible throughout the year. In *Roots Demystified*, author Robert Kourik describes a single cereal rye plant with a root length that measured 372 miles! The root hairs of this plant measured 6,123 miles for a total length of 6,495 miles! That will certainly increase organic matter! How would you like to have been the graduate student who had to dig up and measure that plant?

The role of living roots in pasture and rangeland is just as significant. In my travels, I see tens of thousands of acres planted with only cool-season species or only warm-season species. This lack of diversity is not optimal in terms of maintaining living roots in the soil for the long span of the year. There is a reason most natural rangeland ecosystems have both cool- and warm-season broadleaves and grasses. The ecosystem is healthy only if both are present.

Principle Five: Integrate Animals

The fifth principle is to keep animals present in the agricultural landscape. Another tragic flaw of the current production model is the removal of animals from the landscape. Take a look back at how our grandparents farmed a century ago. Nearly every farm had beef or dairy, along with hogs and poultry. Horses were used as draft animals. Today we have moved the poultry and hogs into confinement buildings, the beef onto feedlots, and the dairy into very large confined operations. In many parts of the world, one can drive for hundreds of miles without seeing a fence, let alone an animal.

What difference does this make? To answer that we must understand how soils were formed. Centuries ago, tens of millions of bison, elk, deer, and other ruminants roamed the North American continent. These ruminants took a bite of a plant here and another there, causing those plants to release root exudates in order to attract biology that supplied the nutrients needed for regrowth. The presence of predators kept the herds of ruminants on the move, and they often did not return to the same spot for long periods of time. The plants thus had ample time to fully recover, all while pumping massive amounts of carbon into the soil. (As noted previously, a plant that has been grazed will photosynthesize more and pump much more liquid carbon into the soil compared to a plant that has not been grazed.) Add to this the myriad of insects, birds, and other wildlife that also lived in these environments, and it all added up to a very healthy, optimally functioning ecosystem.

Today, with grazing animals almost entirely removed from the world's grasslands, there is much less carbon cycled through the system. There are those who blame cattle for climate change. That viewpoint is too simplistic; it does not take into account the larger picture of how ecosystems function. The best-proven way to transfer massive amounts of carbon dioxide out of the atmosphere and into the soil is by maintaining a landscape that includes grazing animals. It is not the cattle that are the problem, it is our management of them! I thoroughly enjoy debating with vegetarians and vegans as to the importance of animals on the landscape. My contention is that if they are truly concerned about the health of ecosystems, they have to recognize the benefits that grazing ruminants provide, even if they choose not to partake in eating meat. One of the best presentations of this argument is put forward in the book *Defending Beef* by Nicolette Hahn Niman.

Integrating multiple species of animals throughout Brown's Ranch has led to much larger amounts of carbon in our ecosystem. This has not only improved soil health, it has also significantly increased our profitability. I speak to hundreds of farm families every year who lament the fact that they are not making a profit. When I ask them about their model of production, I usually discover that they do not run any livestock on their land. I encourage all operators to take advantage of the many benefits that animals offer.

Eight

Growing Biological Primers

Planting cover crops is a key step in transforming dirt into soil. In this chapter, I describe how we use cover crops on Brown's Ranch and share what I've learned over the years from our experience and from other farmers and ranchers working in regenerative agriculture.

Although I first started planting cover crops more than twenty years ago, I didn't think of them as cover crops back then. I was just seeding crops to serve as livestock feed. Even now, I really don't like to use the term *cover crops*. I prefer to call these crops *biological primers* because they do so much more than just cover the soil. For simplicity's sake, though, I will refer to them as cover crops.

If you raise livestock and manage cropland, cover crops are an absolute no-brainer because livestock can help convert the covers to dollars quickly. If you do not have livestock, you should still plant cover crops for a host of reasons, including putting more carbon into the soil, feeding biology, protecting the soil from erosion, and, of course, improving profitability!

Cover Crops Cycle Carbon

What cover crops can, and will, do is increase the amount of carbon in your cropland fields. In chapter 3 I explained how important carbon is and how plants, through photosynthesis, pump liquid carbon into the system. The more leaf area there is in a field, the more sunlight will be "caught," and the more photosynthesis will occur. Dr. Christine Jones calls this *photosynthetic capacity*, and it's another reason why seeding multispecies cover crops is an even better choice than seeding a field to only one or two species. The varying leaf sizes and shapes in a multispecies planting, along with the larger range in plant heights, will result in more sunlight-encountering leaf area, thus pumping more carbon into the soil.

Dr. Jones defines *photosynthetic rate* as how fast a plant can convert light energy into sugars. Many factors affect this rate, including moisture, temperature, light intensity, carbon demand placed on the plant by soil microorganisms, and the presence of mycorrhizal fungi.

The higher the photosynthetic capacity and photosynthetic rate of crop and pasture plants, the healthier the soil ecosystem and the more rapid the building of new topsoil. Remarkably, some plant species can pump up to 70 percent of the carbon they capture through photosynthesis into soil in the form of root exudates. In addition to providing a carbon source, root exudates support free-living and associative nitrogen-fixing bacteria. As soil carbon levels increase, soil structure improves and the conditions for biological nitrogen fixation are enhanced. Needless to say, the availability of those nutrients can lead to much greater profitability.

I often think about this cycle as I drive through agricultural regions, past seemingly endless crop fields left fallow after harvest. To make matters worse, the current practice of many producers is to spray fields with herbicide after harvest so no "volunteer" grain will sprout and grow! Most conventional cropland fields are left bare for six to nine months of the year. And if no green plants are present, there will be no photosynthesis, no conversion of sunlight into stored energy, no capture of carbon dioxide and cycling of carbon in the soil. These practices are fighting against nature every step of the way.

Unfortunately, in many cropland (and pasture) soils, soil life has been decimated by prior farming practices. Tillage, synthetic fertilizers, lack of diversity in the crop rotation, and the outright absence of growing plants for much of the year all contribute to a dysfunctional soil food web. In pasture land, overgrazing, undergrazing, and a lack of diversity can lead to the same condition. In order to have both the numbers and diversity of soil biology needed to perform the tasks necessary to ensure a healthy soil ecosystem, we must have diversity of plant species. Another reason to grow multispecies cover crops.

What a different world we would be living in if we could convince all farmers to plant a cover crop after the grain harvest whenever possible. This one simple, easy-to-do practice would significantly raise the collective photosynthetic capacity of our agricultural land, and that would go a long, long way toward healing our planet!

What Is Your Resource Concern?

People often ask me how I decide which cover crop species to seed in a mix. In order to answer that, I must first ask and answer the question, "What is my resource concern?" In other words, what am I trying to accomplish by planting this cover crop? Do I want to improve the organic matter level of the field? Do I want to improve water infiltration? Do I need to increase species diversity? Improve nutrient cycling (i.e., reduce synthetic fertilizer use)? Control weeds? Manage pests? Address salinity issues? Provide wildlife habitat? Attract pollinators? Feed livestock? And the list goes on. The beauty of it is that cover cropping, practiced correctly, can address every one of your resource concerns.

I often hear producers say that they tried cover crops, but they did not work. In response, I ask them what their resource concern was. Usually that question draws a blank look from the producer. In other words, they seeded a cover crop without first thinking about what they wanted to achieve, which meant they had no logical basis for deciding what species to use. Often, they just seeded what was easily available. The result is usually not good.

In terms of learning what types of cover crops may perform well in your area and deciding which ones will address your resource concerns, it's important to do your homework. There is a wealth of information available online, including the book *Managing Cover Crops Profitably*, which is available for free download on the Sustainable Agriculture Research & Education website, www.sare .org. The book includes descriptions of many common cover crops, where they can be grown, and their growth habits and benefits. I also recommend that you seek out others in your area who are using cover crops and ask them about their experiences. Ask local seed suppliers for advice and attend local field days. And I encourage producers to perform trials on their own operation every year. We try several different species and combinations every year on our ranch. If a species does well, I increase its use the following year. If it fails two years in a row, I do not try it again. It's also important to understand the seasonality of the cover crops you are considering growing. For example, do not plant barley in North Dakota in July. Do not plant millet in North Dakota in April. I often think of this when I see many farmers in the northern plains planting corn in March and April. Last time I checked, corn was still a warm-season grass!

Let's take a look at organic matter as a resource concern. Organic matter level is not the only key indicator in determining soil function; however, it is one of the foundations of a healthy soil. It is important to realize that organic matter levels fluctuate according to climate conditions and management. The definition of *organic matter* is matter that has come from recently living organisms. It is capable of decay, or a product of decay, or is composed of organic compounds. It is the flow of carbon energy through living organisms carrying out their metabolic processes, which create the organo-mineral complexes that entomb or coat the sand, silt, and clay particles.

I have never been on a farm or ranch, including my own, that is not degraded. If you search the archives where you live, you can get a good idea of what soil organic matter levels were a century or more ago. If the organic matter in your soils has dropped as much as they have in my area (from over 7 percent to around 2 percent), you do not have properly functioning nutrient or water cycles on your land. Farms with low soil organic matter must rely on synthetic inputs to do the jobs that nature originally did for free. As organic matter levels rise, and we provide a habitat for soil biology, the amount of available nutrients in the soil increases dramatically. I did some calculations, and at the time of writing, for every 1 percent increase in organic matter, there is the equivalent of about $750 per acre's worth of nitrogen, phosphorus, potassium, and sulfur inputs. Realize that you must first have the biology to cycle these nutrients. Management is key to achieving that, and when you do have the levels needed, you will be able to significantly cut inputs, thus improving profitability.

Approximately two-thirds of the soil organic matter increase will come from establishing roots in your soil. It's critical to put as much root mass as possible into the soil, from the surface all the way down into the subsoil. These roots pump the liquid carbon to feed the biology that is critical to soil function. Cover crops such as sorghum/sudangrass, cereal rye, annual ryegrass, phacelia, and red clover, among others, will produce large amounts of root mass to achieve this. One cover crop blend I like to use is a warm-season blend of sorghum/sudangrass, pearl millet, cowpeas, mung beans, annual sweetclover, sunflowers, kale, daikon radish, buckwheat, and safflower. This mix gives us a good variety of different root types and rooting depths to fill the soil profile, thus increasing organic matter. The different leaf shapes maximize solar energy collection, and the variety of flowering species attracts beneficial insects. Keep in mind, though, that while these species work on our operation they may not work on yours. You will never know unless you try!

Determining Seeding Rate

One of the questions I hear most frequently is, "How do I determine what percentage of each species I should put in the mix?" Let me run through the example of the mix I stated above. My resource concern is increasing organic matter, so I want the highest percentage of the mix to be sorghum/sudangrass and pearl millet. Those species will put the most root biomass in the soil. I then want to add some legumes in order to fix nitrogen. Cowpeas, mung beans, and annual sweetclover work well in a summer mix in my environment. Sunflowers have a long taproot that will bring up nutrients from deep in the soil profile. Buckwheat secretes root exudates that attract biology, which will make phosphorus available, and its flowers attract pollinators. I add safflower to the mix because snow is a big source of our moisture and I need an upright plant that will catch snow. Cattle and sheep do not relish safflower, which also is in its favor for serving as a snow catcher. I include kale because it is a prolific producer of high-quality forage. Daikon radish provides a deep taproot that serves as a nitrogen storage sink, soaking up nitrogen to be released the following spring when the tubers decay. In table 8.1, I present an example of the rates I use for this blend, but again, what works in my situation may not work the same in yours.

You may be wondering, how does one decide how many seeds to plant per acre? I wish there was a solid answer to that question, but the truth is it depends on the growth habits of the species in the mix. Do they have an upright growth habit such as a small grain or sorghum/sudangrass, or will they stool out and cover a larger area such as some of the forage brassicas, vetches, and clovers? Nothing beats experience here. The best sources of advice on seeding rates are producers who have been seeding covers for years and the staff at seed companies that specialize in cover crops.

The next two questions growers commonly ask me are, "Won't the small seed settle to the bottom of the drill box?" and "How do I set the drill?" As long as you don't fill an air seeder and head out

Table 8.1. Example of Seeding Rates for a Ten-Way Cover Crop Blend

Species	Pounds of seeds per acre	Seeds per pound	Total seed count
Sorghum/sudangrass	12	18,000	216,000
Pearl millet	2	80,000	160,000
Cowpeas	10	4,100	41,000
Mung beans	5	12,000	60,000
Annual sweetclover	1	70,000	70,000
Sunflower	0.5	8,000	4,000
Buckwheat	2	18,000	36,000
Safflower	1	15,000	15,000
Kale	0.5	175,000	87,500
Daikon radish	1	25,000	25,000
Totals	35	425,100	714,500

to seed a thousand acres at one time, the seed is not going to settle much. Besides, even if it does settle a bit, it is a cover crop; it doesn't have to be perfect!

As for setting the drill, the best method is to start by jacking up the drill and measuring the circumference of the drive wheel. Then fasten bags over the openings of two or three seed tubes to collect the seed that will be dispensed. Turn the drive wheel the correct number of revolutions to equal one hundred feet. Use a gram scale to weigh the seed that is dispensed, and divide that by the number of tubes you collected seed from to get the average weight of seed dispensed per tube. Use the following formula to calculate total pounds of seed per acre:

$$\frac{43{,}560 \ (\text{ft.}^2/\text{acre}) \times \text{lb. seed dispensed (lb. seed/tube} \times \text{total tubes})}{\text{Drill width (ft.)} \times 1.1} = \text{lb. seed/acre}$$

The factor 1.1 is included to adjust for tractor tire slippage in the field. This calculation does not take long to do and really works well.

Improving Water Availability

Is the availability of water your resource concern? Surviving a drought, for example? By improving soil structure, we also increase the ability of our soil to both infiltrate and store water. Soil holds onto water via a capillary film around each soil particle. The more soil particles (aggregates), the greater the water holding capacity. One of the biggest resource concerns in the Upper Midwest is sheet-wash erosion and flooding, which are just other ways of describing poor water infiltration into the soil. When we purchased the farm in 1991, the infiltration rate on our cropland was only one half inch per hour. When a big storm came along, dumping two or three inches of rain, most of the water left the farm in a hurry, usually taking a bunch of topsoil with it. By 2009, the infiltration rate had risen to more than *ten inches* per hour thanks to well-aggregated soils due to mycorrhizal fungi and soil biology. The soil could absorb large volumes of water, which is exactly what happened on June 15, 2009. It started raining at 5:30 in the evening and kept raining, and raining. When the storm was over the following morning, we had received 13.6 inches of rain in twenty-two hours, but the vast majority had infiltrated into the soil. Jay Fuhrer visited the farm that day to see how our soils had held up, and he said that you could have driven a vehicle across the fields without making a rut.

In 2015, we had a researcher film an infiltration test on our cropland in which an inch of water infiltrated in nine seconds. The second inch in sixteen seconds. That is a huge improvement from a half-inch an hour!

How much rainfall you get is not important; what is important is how much rainfall can infiltrate the soil.

It puzzles me how many producers choose tillage or tile drainage to solve water infiltration problems. By doing so, they are only treating a symptom not solving the problem. My good friend David Brandt's land is proof that farmers do not need either tillage equipment or tile

Plate 1. Soil like chocolate cake—rich, dark, and well aggregated—supports vigorous roots and abundant soil biology. This cropland field, planted here to red clover, had been no-till for fifteen years.

Plate 2. Life! This is a handful of soil from one of my no-till cropland fields. When we bought our farm in 1991, there were no earthworms in any of the cropland fields. Now they are plentiful.

Plate 3. Diverse perennial pastures like this one are where I learned about "nature's way" of maintaining a healthy, functioning ecosystem.

Plate 4. Notice how clear the water is in this seasonal waterway flowing through a native pasture. No soil erosion problem here!

Plate 5. The diversity of plant species and healthy soils in our perennial pastures allows them to produce well even during a very dry year. This is a great example of a healthy, functioning ecosystem.

Plate 6. Our animals also thrive on cropland like this that has been seeded back to perennial forages.

Plate 7. When a hailstorm like this one forms over Brown's Ranch, we are fully exposed. Our farming practices help us to be resilient even in the face of extreme weather events.

Plate 8. Armoring the soil is one way to be resilient. The armor is the residue from a previous cover crop and a cash grain crop is growing through the armor.

Plate 9. Notice the uniform aggregation of this soil from a crop field at Brown's Ranch. Mycorrhizal fungi secrete glomalin to build soil aggregates.

Plate 10. After only two years of growing cover crops, the soil from this old hayfield shows almost miraculous improvement.

Plate 11. This residue was left after our cattle grazed a warm-season cover crop in late fall or early winter. We made sure at least 65 percent of the aboveground biomass was left after grazing. The armor prevents evaporation and wind erosion, and inhibits weed growth.

Plate 12. Here is what 700,000 pounds live weight per acre looks like grazing on diverse perennial forage.

Plate 13. These content cow/calf pairs are grazing a perennial pasture of expired Conservation Reserve Program acreage.

Plate 14. We use rubber truck tires as water tanks. Shallowly buried pipelines supply the tanks, which are set underneath the fence line that separates two paddocks.

Plate 15. We set out bales of hay for winter grazing. By moving the bales all at once in the fall, we save on labor costs from moving feed daily in difficult winter conditions. This practice also improves soil quality.

Plate 16. Wildflowers in our perennial pastures attract pollinators and other beneficial insects.

Plate 17. We also include pollinator plants such as sunflowers in our cover crop mixes. This practice helps to control pest insects.

Plate 18. Every farm should include pollinator strips like this one on Brown's Ranch. This is a warm-season mix of big bluestem, Indiangrass, switchgrass, chicory, white clover, and crimson clover.

Plate 19. This hugelkultur planting in our vegetable garden is a diverse polyculture mix of corn, beans, squash, sunflowers, and flowers for pollinators.

Plate 20. We don't till for any crop, not even potatoes. Instead we simply place the potatoes on the soil surface and then unroll second cutting alfalfa hay over the top.

Plate 21. When it's time to harvest the potatoes, we simply roll back the hay. No digging required!

Plate 22. This fall-seeded cover crop of oats, peas, lentils, and daikon radish is "capturing" sunlight. I never pass up the opportunity to cycle carbon, even when frost may hit in early September.

Plate 23. A cover crop sown in the fall helps provide weed control the following spring.

Plate 24. This fence line comparison shows the difference between a pasture belonging to a neighbor that has been grazed all season long, and one of my pastures grazed using Holistic Planned Grazing.

Plate 25. Notice the variety of seedlings emerging? This is what it looks like when you seed a field with a diverse polyculture cover crop.

Plate 26. Regeneration in action: This diverse cover crop mix of sorghum/sudangrass, pearl millet, cowpeas, mung beans, forage brassica, buckwheat, and safflower seeded into an existing alfalfa stand is regenerating the health of the soil—no inputs required.

Plate 27. Our ranch sign advertises our name—and our farming principles.

Plate 28. We use a concession trailer as our stand for selling our Nourished by Nature products at farmers markets.

Plate 29. The Batt-Latch gate at lower right opened up automatically, allowing the cattle to move through to a new paddock.

Plate 30. This group of grass-finished beef animals is grazing a mix of sorghum/sudangrass, pearl millet, cowpeas, mung beans, guar, forage brassica, sunflowers, and daikon radish.

Plate 31. This is the first eggmobile we built: a livestock trailer to which we added roosts, nest boxes, and a waterer.

Plate 32. Hogs thrive in pasture, too. These hogs are grazing an annual cover crop mix.

Plate 33. The difference between conventional agriculture and regenerative agriculture: A neighbor applies synthetic fertilizer while the Brown's Ranch "fertilizer crew" applies natural fertility.

Plate 34. Part of our two-hundred-year plan is to plant more perennial crops. This orchard of fruit trees is just getting started, but imagine it thirty years from now. And the understory here is not weeds—it's a mix of plants that attract pollinator and predator insects.

A Better Way to Fight Pests

From what I observe, crop pests are a resource concern that most producers are having more issues with than they did in the past. Their answer is to either spray with pesticides or use genetically modified crops. I used to do those things, too, until I added diverse cover crops into my crop rotation. It didn't take too many years until I noticed that I no longer had issues with pests. In fact, I have not used a pesticide since before the turn of the century, except for seed treatment on corn, which I discontinued in 2010. Why was I able to do this? I attribute it to growing diverse cover crops and including flowering species in those mixes. By doing so, I was attracting and providing a home for the predators. An expansion of this is to plant perennial pollinator and predator strips, which I discussed in chapter 3.

drainage to solve water issues. In his area of Carroll, Ohio, annual rainfall is well over forty inches a year, and the soils have very high clay content. David's soil, however, is unlike the majority of soil in the area. Due to his use of no-till and cover crops, his soil is well aggregated and has the ability to allow infiltratration and move the water throughout the profile. Dave uses a variety of cover crops that have the ability to build soil aggregates and poke holes through the clay, such as daikon radishes, sunflowers, alfalfa, deep-rooted clover, rye, ryegrass, yellow blossom sweetclover, and sorghum / sudangrass.

Another critical factor in water-holding capacity is organic matter. For every 1 percent increase in organic matter, we can store between

17,000 and 25,000 gallons of water per acre. To use my operation as an example, I went from less than 2 percent organic matter in 1991 to well over 6 percent in 2017. In 1991 my soils could store approximately 40,000 gallons of water per acre, while in 2017 they were able to store well over 100,000 gallons per acre. That difference is huge—and critical to the success of our operation—when timely rains do not come.

A couple of years ago, I was touring a ranch in California where there was a lot of talk about the "drought" that they were coping with and how it had lasted five years. The grasses and forbs in the pastures were all annuals and had only a few inches of growth on them. Digging up a spade full of soil revealed very shallow root systems and poor aggregation. Obviously, the soil was low in organic matter. I asked the owner how much rain had fallen that year. "Only thirty-two inches!" she exclaimed. This was a perfect example of how we can "create" our own drought. I firmly believe that we, as producers, have the ability to make our soils and thus our operations much more, or less in this case, resilient to wide swings in moisture.

When I make the statement that the amount of rainfall an area receives is not relevant, I get a lot of disgusted looks—especially from farmers in drier climates. It is true though. What does matter is how much rainfall infiltrates the soil and then can be stored via organic matter. This is called *effective rainfall*.

I once gave a presentation in a part of Arkansas where the average annual rainfall is fifty-four inches per year. Some producers in that area were putting another fifty inches of irrigated water on their fields to produce 175 bushels of corn, which equates to a water use efficiency of only 1.7 bushels of corn for every inch of water. Compare that to my average yield of 127 bushels of corn from 15.7 inches of annual precipitation for a water-use efficiency of 8.08 bushels of corn per inch of moisture. Those farmers in Arkansas should easily be able to yield 175 bushels of corn per acre without any irrigated water, if only they would address the issues that are limiting water infiltration and storage.

Addressing Nutrient-Cycling Issues

I have visited many farmers who have converted unproductive cropland to perennial forages only to find that those perennial forage stands were not productive either. The reason this occurred is that they did not address nutrient-cycling issues first. They now have an unproductive stand of perennials. Let me describe one way to address this type of resource concern.

On our ranch, we've seen excellent results by planting fall-seeded biennials such as cereal rye, winter triticale, and hairy vetch directly into the perennial stand. We do not terminate the perennials by spraying herbicide. The following spring, we allow the biennials to grow until they are starting to flower, and then we graze them. We limit the livestock to consuming about 35 percent of the above-ground biomass. They trample the rest, leaving a nice thick layer of armor (residue) on the soil surface.

Then we interseed a predominantly warm-season cover crop mix. If at all possible, I advise doing this without mechanical tillage because of the harm tillage causes (if you've forgotten about this, return to chapter 7). Most of the no-till drills on the market today will be able to seed directly into this type of sod. But I do not want to mislead you; seeding annuals into an existing perennial stand is not easy. Good seed-to-soil contact is critical. We make sure there is enough downpressure in the drills in order to cut through the surface residue, get the seed to the desired depth, hold it in place, and cover the seed trench. The majority of failures I see in seeding cover crops into existing perennial paddocks are a failure to achieve good seed-to-soil contact and then cover that seed with just the right amount of soil to ensure good germination.

I like to use sorghum/sudangrass, pearl millet, kale, daikon radish, sweetclover, sunflower, buckwheat, and safflower. I let my livestock graze this mix late in the season, thus allowing the covers to grow to their full potential. We repeat the process the following year but with different species in the cover crop blend. In the spring I seed a mix of barley, oats, peas, and kale. Again, this mix is seeded directly into the sod, early, before the perennials start to grow. Our livestock graze this mix in late June. Keep in mind that the perennials are still alive and grow as well but are somewhat shaded by the covers. Once grazing of this forage is complete, I seed a warm-season mix directly into the residue. We do not graze this mix until late fall or winter.

The third year, I seed oats along with whatever perennial grasses, forbs, and legumes I may want in the long-term perennial stand. I have had very good results improving both nutrient cycling and soil quality by using this method.

It is important to realize that in a healthy soil the microbes themselves also store and cycle nutrients. Soils contain an estimated two to three million *species* of bacteria, of which only 2–5 percent have been described or named. The reproductive potential of bacteria is incomprehensible. A single bacterium allowed to divide every hour would yield seventeen million cells in only twenty-four hours. Some species can double their population in as few as twenty minutes. These masses of bacterial cells consist of up to 60 percent nitrogen. Thus, they are a huge nitrogen pool in healthy soils.

Another important note about nutrient cycling: Only about 40 percent of the synthetic nutrients applied in a given year on croplands are actually taken up by the plants that year. The remainder stays in the soil and, more often than not, especially in the case of nitrogen, is lost due to leaching out of the soil profile. The best way to prevent this loss is by growing a cover crop. Biology converts that synthetic nitrogen fertilizer into inorganic forms of nitrogen (nitrate and ammonia) that can be taken up by the cover crops, where it is "stored" in a living plant. Once that plant runs its life cycle, the nutrients will be cycled. Why would a producer want to write a check for synthetic fertilizers and then not store it on their land?

Developing a Crop Rotation Plan

Many producers tell me they cannot fit cover crops into their rotation. I tell them that they need to make covers a priority—change the rotation! An easy way to do this is with fall-seeded biennials. One mix that is a no-brainer in my operation is a simple combination of cereal rye, hairy vetch, and daikon radish. The cereal rye has tremendous root mass, which improves soil structure and increases organic matter. The hairy vetch, being a legume, hosts rhizobia that can convert atmospheric nitrogen into useable forms. The radish will not overwinter but will store and cycle excess nitrogen. This cover crop gives me a lot of options. I can combine it, I can graze it, or I can terminate it and then seed a cash crop into that residue.

The ways in which we fix cover crops into our crop rotation on Brown's Ranch may not work for you, but I explain my methods and choices as a way of helping you think through what will work on your operation. We have about 2,000 acres of cropland on our ranch, which we use for a variety of cash grains, cover crops, and forage. One important goal we strive for is to have living roots in the soil for as long as possible on these cropland acres throughout the year.

As I described in part one, when Shelly and I began farming, I started to diversify a little bit right away from the routine of spring wheat, oats, and barley that Shelly's parents had grown year after year. What they did was the norm for most farmers in North Dakota at that time, and even today, most producers grow only a few cash crops. I chose to add peas and hairy vetch, which are cool-season broadleaves, and corn, millet, and sorghum/sudangrass, which are warm-season grasses. Flax and sunflowers were added to fill the warm-season broadleaf component. This broadened my crop rotation to include species from the four crop types: warm- and cool-season broadleaves and warm- and cool-season grasses. Diversity is the key.

How do I develop a crop rotation with so many cash crops? The answer is simple: I call it organized chaos. I do not have a set crop rotation. To do so signifies repetition, and repetition is a setup for failure.

Look at natural ecosystems—are they repetitive? No! In nature, different plant species express themselves according to the conditions. Moisture, temperature, sunlight, humidity, and a myriad of other variables dictate which plants and animals, including insects and even soil microbes will thrive that year. Once you begin to understand the power of diversity, you'll be motivated to diversify your crop rotation.

My thought process is that on any single field I will try to plant at least three of the four crop types in a four-year time period. All four is the ideal, but some years I do not meet that goal. Many people tell me, "That's no rotation!" But as I explained, if I set up a fixed rotation, nature will figure it out. Pests that attack crops each year will acclimate to the set rotation and will feast as a result. The ever-prevalent corn/soybean rotation, which proliferates the corn borer, corn rootworm, and soybean cyst nematodes, is a perfect example of this. Pests on my ranch have no clue what is coming next and thus cannot gain a foothold.

Managing Synthetic Inputs

The first thing most producers ask me about my cover crop management is whether and how I use an herbicide. Like any input, I try to use it as judiciously as possible, which typically means only one herbicide pass every two to three years. I have gone as many as five years without it, but it's critical to keep an eye on perennial invasive species, especially if no forage or crop residue is processed (baled or green chopped) and removed. Rarely do I use a post-emergent herbicide. I do not want any synthetic chemicals sprayed on crops that go for human food or livestock feed. I don't offer any specific recommendations about which herbicides to use. There are just too many formulations and variations for me to be able to do that.

One criticism I often receive is that I still occasionally resort to using an herbicide to control weeds. Yes, I do, and it bothers me to do so, but it does not bother me nearly as much as using a tillage pass would! In my opinion, tillage is much more destructive to a soil ecosystem then an occasional herbicide pass. I am not making an excuse, and I am working diligently to end my use of herbicides, but I will not use tillage in its place. Of the hundreds of farms and ranches all over North America that I visit each year, including dozens of organic farms and ranches, not one of those farms has soil quality that compares favorably to the soil on my ranch. On the organic farms, they use no herbicides, but they do till, and that simply destroys soil structure and function.

I have noticed, both on my farm and visiting others, that as soil health advances, weed pressure declines. This seems to be particularly true when the fungal to bacterial ratio in the soil approaches 1:1. Recent research by Dr. David Johnson of New Mexico State University indicates that the ratio between fungi and bacteria in the soil is critical to a plant's productivity in healthy agricultural systems and thus to a plant's efficiency in nutrient uptake. In a forest ecosystem, fungi are dominant, with the fungi:bacteria ratio at 100:1 or higher. In bare soil, such as a tilled field, the ratio is reversed to 1:100; bacteria dominate. Because one of the purposes of regenerative agriculture is to restore degraded land to health, Dr. Johnson contends that the most effective fungi:bacteria ratio for regenerative systems, depending on the crop, is between 1:1 and 5:1. This is important because moving soil into this balance involves more than simply reducing synthetic fertilizers, herbicides, and pesticides (as in organic farming systems). It means finding a way to promote fungal activity in the soil, which, of course, means following the five principles of a healthy soil ecosystem that I outlined in chapter 7.

Dr. Johnson's research shows that the fungal component of the soil is the most important factor to most plants early in life. It is much, much more significant than nitrogen, phosphorus, potassium, or even organic matter. In fact, he has documented that some plants secrete up to 96 percent of the carbon that they cycle in order to feed soil fungi and biology. Wow! How powerful is that? Yet how many of us, as producers, agronomists, or gardeners, are aware of this, let alone focusing on it?

The intelligence of the natural world is working for the benefit of the ecosystem which, if we allow her to, produces nutrient-dense food. Diverse plant roots, sending diverse signals to bacteria and fungi, make diverse minerals available as needed, all working together symbiotically to transfer nutrients into the body of plants and then into us, in an elegant, efficient, and vastly sophisticated, interdependent web of life, billions of years in the making. A plant strives for balance, health, diversity, and renewal, tightly bound to the specifics of a particular place, while honoring universal biological principles. Powered by the sun, energized by liquid carbon, and supported by a complex array of microbes, a plant is a miracle of nature, sustained by an intricate network of intimate, time-honed relationships.

Dr. Johnson and his wife, Hui-Chun Su, have developed a static compost system that allows the fungal community to be undisturbed and dominate. It's aerobic, always having access to oxygen, and it's a vermicomposting process, allowing nature's worms, the ultimate degrader, to have their way with the quality of the finished compost. The compost is left in the bioreactor for one year in warm climates, longer in cold climates. This resting period allows the compost time to develop the species diversity that is found in healthy soil ecosystems. A slurry made from the mature compost can then be used to "inoculate" seeds prior to planting, or as little as 400 pounds of compost per acre can be spread on the soil to introduce the biology. (It should be noted that getting the compost into the soil is preferred as compared to applying it to the surface.) Results from this system are very encouraging. I believe this system will allow us to regenerate degraded soils at a much faster rate than we ever thought possible.

A good friend, Vincent Mina of Maui, Hawaii, is using a somewhat similar system called Korean Natural Farming on his sprout farm. The results are outstanding. I feel that both of these systems have merit, and I will be exploring their possibilities on our ranch in the coming years. It is my hope and goal that as I continue to advance soil health, weeds will become a nonissue, and I will be able to phase out all use of herbicides.

How about fertilizer? As I was writing this book, I discussed soil fertility with Dr. Christine Jones. I asked her, "How many places in the world are there where true nutrient deficiencies in the soil make the land unsuitable for profitable production agriculture?" "Very, very few," she replied. The nutrients are there; soil biology is what is needed to make those nutrients available. It is possible to reduce or even eliminate your usage of fertilizers, but, as I noted previously, the weaning process should be gradual. Using the Haney Soil Test as a guide is a great way to move in the right direction—away from synthetic fertilizer.

I started with one simple step: I added field peas, a legume, to my rotation in order to take advantage of all of that "free" nitrogen in the atmosphere. Remember, above every acre there is approximately 32,000 tons of atmospheric nitrogen. Why would any farmer purchase synthetic nitrogen fertilizer, rather than "harvest" their own? I noticed an immediate improvement in soil health and subsequent crops after adding peas to my crop rotation.

Whenever I harvest an early crop such as oats, barley, or peas, I follow on afterward by seeding a rye/hairy vetch blend along with a small amount of daikon radish. The radish scavenges nitrogen, storing it for subsequent crops.

Cereal rye or winter triticale mixed with hairy vetch are a mainstay in my cropping system. I seed several hundred acres of these fall-seeded biennials each year. They give me a lot of options:

- I can combine the crop for a cash grain crop. This particular mix has been my most profitable cash crop for the past nine years.
- I can graze this mix in the spring with virtually any class of livestock, and, if I use good grazing management, I can graze it multiple times. It is a great mix to calve cows on.

- I can graze it only once and then allow it to regrow and mature, after which I combine it for grain.
- I can cut it for forage, although this is something I would rather not do because I do not want to remove that much carbon from a single field.
- I can terminate the mix with an herbicide and seed another cash crop into the residue. I rarely do this, but it is an option.

The hairy vetch seed that I have today originates from the first seed I bought back in 1994. In essence, by saving seed for more than two decades, I have developed my own variety that is uniquely suited to the environmental conditions of my ranch. This mix has never failed me. I always get some production out of it; thus, it is like an insurance policy to me. I recommend saving seed to all producers; it's a great way to gain some peace of mind.

One of the most productive cash crops we grow is oats, which we often mix with field peas—a legume and a grass, working together in beautiful symbiosis, just as nature intended. We combine this as a mix and then either sell it in cover crop blends or feed it to the pigs and chickens. This mix also leaves us with the option of grazing or haying it if we feel we will need the forage.

When we want to harvest a straight oat crop for grain, we seed clovers along with it. We use lower-growing clovers such as crimson or subterranean to ensure that the oats will be taller, thus allowing us to straight cut when we combine. (*Straight cut* means that the grain crop is not swathed first.) The clover cycles nitrogen for the oats and the subsequent crop. It also provides grazing once the oats are combined. Mycorrhizal fungi really like oats and will proliferate in an oat field. I have also been saving my own oat seed for decades to ensure that it is acclimated to my soils.

I usually plant a legume cash crop the year preceding corn, to cycle nitrogen into the soil. If the peas are harvested early, I follow with a warm-season cover crop mix of sorghum/sudangrass, buckwheat, cowpeas, mung beans, and guar. All of these will cycle more carbon and will winterkill. The sorghum/sudangrass provides a nice thick coat of armor, which prevents weeds from becoming an issue in the flowing corn crop.

Like many of my neighbors, I grow corn. Unlike my neighbors, I rarely grow it as a monoculture. I plant the corn seed and then, about three days later, I go in with my grain drill and seed a mix of clovers and hairy vetch directly into the field already planted with corn. The corn germinates and emerges first and gets going ahead of the legumes. The legumes cycle nitrogen, provide weed control, attract pollinators, and provide late fall or winter grazing.

Many producers in higher rainfall areas have good results with broadcasting the clover/vetch mix into newly emerging corn. I have had marginal results with this technique in my environment. However, don't be afraid to experiment. What will work best for you will depend on your farm's unique conditions.

Following a corn crop, I like to seed a cold-tolerant cash crop such as oats and clovers or barley and clovers. This crop provides early canopy to help prevent weed seeds from germinating. Sunflowers are also a very good option following corn because of their deep taproot; they can thrive if moisture conditions are marginal. Be sure to consider planting covers along with them.

The deep taproots of sunflowers will help cycle deep nutrients. It is also very easy to interseed covers into them. I do not grow sunflowers every year, but when I do, I like to seed a diverse mix of annual warm-season species such as millet, mung beans, guar, buckwheat, and flax into them. I seed this mix within a few days of planting the sunflowers, which are spaced on thirty-inch rows. The warm-season cover crop species will be terminated by frost, thus making harvest of the sunflowers a breeze.

In 2017, I took diverse polyculture cash crops to a new level when I mixed oats, barley, peas, lentils, and flax together. What? Five crops together? That's just nuts, many people told me. Why would I do this? I had good reasons. The oats and barley have fibrous root systems, thus building soil aggregates and increasing organic matter. They also help cycle phosphorus, which are then used by the legumes. Rhizobia become established in the legume (pea and lentil) roots and fix nitrogen for both the legumes and the oats and barley. The flax is in the mix because of the health benefits from the seeds, which are high in omega-3 essential fatty acids. I combined this mix and, despite our unfavorable weather that year, yielded 62 bushels per acre. I considered it a success. The mix will be used either for seed or feed.

Most producers rely on antiquated conventional soil tests to determine the amount of nutrients they have in their soils. They have the misconception that these tests provide accurate data, but that is just not true. Let's look at nitrogen as an example. Nearly every conventional soil test used today measures only the plant-available inorganic (no carbon molecule attached) pools of nitrogen: ammonium and nitrate nitrogen. Yet Russian soil microbiologists showed a century ago that plants can take up organic (carbon molecule attached) pools of nitrogen in the forms of amino acids. Yes, there is direct uptake of organic nitrogen by plant roots. Today we have the technology to measure these organic forms of nitrogen in soil samples. However, conventional soil tests do not. Thus those tests do not account for a large percentage of the soil nitrogen supply that is available to plants. Because they rely on the results and recommendations from conventional soil tests, most producers are overapplying nitrogen and other nutrients such as potassium chloride, noted a 2013 meta-analysis research paper, "The Potassium Paradox: Implications for Soil Fertility, Crop Production and Human Health," from researchers at the University of Illinois.

The overapplication of synthetic nitrogen negatively impacts soil biota, animal health, plant health, human health, and soil aggregation. Also, this salt loading with chemical nitrogen diminishes the soil's ability to self-heal, self-regulate, and self-organize, which in turn impacts the soil's ability to cycle water and nutrients effectively. Not to mention the fact that when nutrients (whether natural or synthetic) are overapplied, the excess either runs off the soil, leaches through the soil, or does both. These excessive unutilized nutrients reach the lowest points of a watershed: lakes, rivers, and bays. The quantities of nitrates and phosphates in our watersheds is staggering. From the Mississippi Delta, to the Great Lakes, to the Chesapeake Bay, to the San Francisco Bay, and all points in between, we are seeing major issues from the overapplication of nutrients. It is a senseless waste for producers to spend all that money on nutrients that are not even retained in the soil. If producers would simply plant cover crops and reduce the amount of applied nutrients, they would save thousands of dollars annually on fertility costs and sequester a majority of the nutrients in the form of cover crops. And once those cover crops (also known as biological primers, nutrient sequesters, or energy transformers) decompose, the nutrients in those plants will be released, making them available for the next crop.

There are no winners when one overapplies fertilizer, with the exception of the fertilizer salesman, of course. One of the biggest advancements in the health of the soils on my ranch, not to mention the health of my pocketbook, occurred when I decreased and eventually eliminated the use of synthetic fertilizers.

A New Way of Soil Testing

Another simple step that most farmers and ranchers can take to reduce their use of synthetic fertilizer would be to simply change the type of soil testing they do. They should switch from conventional tests to the new biologically driven soil test developed by Dr. Rick Haney, a soil scientist who works for the USDA Agricultural Research Service.

Ray Archuleta met Dr. Rick Haney in 2011 when Dr. Haney gave a talk on soil testing at a Texas NRCS employee soil health course. Ray immediately understood that Dr. Haney had figured out an important piece of the puzzle to soil health. Ray recalled an epiphany he had a couple of years prior while visiting Schrack Dairy in Loganton, Pennsylvania. Jim Harbach (the owner) and Ray were walking on one of Jim's long-term no-till corn fields, which was smothered with worm castings. Jim had planted multispecies cover crops on this field for the last two years prior (inspired by a visit to my ranch). He was also applying dairy manure to his fields. Jim's soil and corn looked awesome. This corn was not lacking nitrogen.

Yet Jim had applied another fifty units of synthetic nitrogen fertilizer to his green field of corn. Ray asked Jim, "Why did you apply this nitrogen?" Jim told Ray, "That's what the soil test said to apply." Ray responded, "What did that cost"? Jim's reply: "$50,000 dollars for the whole farm"! In dismay, Ray exclaimed, "Jim, when was the last time you went on a vacation?" Jim did not respond. Ray went on, "This corn did not need the extra nitrogen. You could have paid for the whole family to go to Hawaii for the cost of this nitrogen application." Thinking about how Jim had been misled by the soil-test recommendations, Ray realized something was very wrong with the soil-testing process. From his past experience, Ray knew that conventional soil tests do not work well for estimating nitrogen and other nutrients. The soil tests were giving inaccurate results, and it was costing farmers millions, if not billions, of dollars a year! This overapplication was also severely affecting the health of our planet.

That story illustrates well the reason that a new soil test was needed. Conventional soil tests evaluate only the chemical and physical properties of the soil. Conventional soil tests use caustic, reactive acids, such as nitric acid or sulfuric acid, which do not mimic the interaction of soil nutrients and plant roots. These tests ignore the fact that 90 percent of the nutrient cycle is *biological*. They do not take into account how soil and plant biology function. In other words, how do plant roots extract nutrients from the soil? Plants do this by excreting exudates, hundreds of carbon-based compounds, including sugars, proteins, organic acids, and other water-soluble compounds.

Dr. Rick Haney realized that in order for a soil test to be meaningful, the test had to biomimic the three most common acids (oxalic, malic, and citric acid) emitted by plant roots. The test should also use water as the extract—since it rains water! Rick's approach to soil testing is based on green chemistry, an approach that seeks to mimic nature's chemistry. This passive approach to soil testing allows the soil to gently express the quantities of nutrients available to plants versus the conventional reactive soil test, which uses caustic acids to "force" the soil to release plant nutrients. These caustic chemicals are never present in the soil naturally because plant roots do not produce them. Conventional soil testing used the typical approach in modern agriculture: Let's force the system to behave in the way we think is best, instead of listening to the natural ecosystem.

The Haney Soil Test measures seven parameters related to soil biology:

- Water-extractable organic carbon (WEOC)
- Water-extractable organic nitrogen (WEON)
- Percent microbially active carbon (MAC)
- Inorganic nitrogen and phosphorus levels
- Organic nitrogen and phosphorus levels
- Organic C:N ratio
- One-day CO_2 respiration

All seven parameters are used to arrive at a final soil-health score. By knowing the results of these individual measurements, the Haney Soil Test can determine the amount of nitrogen, phosphorus, and potassium that is available and will be made available during the growing season.

I encourage producers to take soil samples, split the combined sample in half, and send one sample to the lab that they have been using and the second sample to a lab that will run the Haney test. Then, when they receive results, they should fertilize one-half of the field according to the recommendations of the Haney test and one-half of the field according to the recommendations of their regular lab. The results will speak for themselves, and remember, it is profit, not yield, that matters. I have seen the Haney test used on hundreds of thousands of acres with very high accuracy. It has saved producers millions of dollars. For a good example of how it has positively affected the bottom line of a producer, read the case study of Russell Hedrick on page 151.

An Australian Example

My discussion of cover crops would not be complete without telling you about Colin Seis, a sheep rancher in New South Wales, Australia. His story is remarkable.

Colin helped pioneer a system of farming called *pasture cropping*, by which a cash crop, often a cereal, is no-till drilled into a perennial warm-season pasture when the pasture is in its dormant phase. The cereal crop emerges and grows, and then as temperatures rise, the warm-season perennials also start to grow, providing a living understory. After the cash crop has been harvested, the warm-season perennial pasture comes on more fully, and Colin grazes it with his sheep. Not only is he getting two crops from the same acre, the diverse mixture of warm- and cool-season plants creates a carbon-rich resource underground.

Like us, Colin came to pasture cropping by a long journey that began with a disaster on his farm, which is named Winona. In his case, a wildfire in 1979 destroyed nearly the entire farm and sent Colin to the hospital. When he emerged, he knew he had to rethink his whole approach to farming—because he didn't have any money left. That sounds familiar! Colin knew that the farm had been on a downward spiral economically and ecologically for years. Yields were down, as were carbon stocks in the soil. His father had farmed conventionally for decades, including heavy use of superphosphate, which government agronomists encouraged Australian farmers to use in large amounts. For a while, yields rose with this synthetic fertilizer, but trouble soon followed in the form of declining soil health, rising salinity, and weed and pest outbreaks as ecosystem functions weakened.

Step by step, Colin brought the farm back to health. He rebuilt the burned infrastructure. He refused to use any pesticides (whose costs kept rising). He explored ways to bring back native grasses in abundance. These grasses kept wanting to come back anyway, Colin noticed, so why not encourage them? Colin and his neighbor Darryl Cluff frequently discussed farming over a glass of beer. Between them, they came up with a crazy idea: pasture cropping. Would it be possible to no-till cool-season cereals into dormant warm-season pasture, they wondered? Would it work?

The answer is yes! Cool-season plants (called C3) and warm-season plants (called C4) differ in the leaf anatomy and enzymes used to carry out photosynthesis. C3 plants are generally higher in protein and energy. C4 plants are more efficient at gathering carbon dioxide and utilizing nitrogen. They also use less water to make dry matter. Pasture cropping takes advantage of the ecological relationships between C3 and C4 plants, including dormancy, growth cycles, water needs, nutrient requirements, and diverse associations with soil biology.

This is exactly how nature operates, of course—annuals, perennials, cool-season, warm-season, and animals all working together in organized chaos.

Colin has addressed multiple resource concerns by pasture cropping and has reaped substantial economic benefits as well. He can run more sheep, and the quality of the wool has risen. Winona is nearly all "native" grassland now, with over fifty different species of pasture plants. Crop yields are strong. Nutrient cycling has improved. And perhaps most importantly, soil carbon levels have more than doubled and soil water-holding capacity has increased significantly since he began pasture cropping!

I see real possibilities for pasture cropping in many areas around the world. In the United States, for example, the south central and southeastern states are dominated by warm-season species. Cool-season cash crops could be seeded directly into those perennials in the fall.

Like Colin, my experiences after the crop disaster years led me to find out what works on my operation. I wanted to make sure we were resilient in case another disaster hit. Let me emphasize that cover crops are not a cure-all. They are a piece of the larger puzzle, albeit a very important one. Ask yourself what your resource concern is and then address it. Living plants can regenerate soils. As Ray Archuleta likes to say, "Plant and soil are one!"

Nine

Will It Work on Your Farm?

During a presentation or farm tour, one of the questions I hear most often is "Your method may work in North Dakota, but will it work where I live?" The questioner usually sounds skeptical—as in, *It'll never work on my place.*

When I press for more information about their operation, the farmer or rancher will usually have an excuse ready: It won't work because their farm has too much clay, or too little; it's too dry, too wet, too low, too high, too rocky, too compacted, and so on. My response is always the same: Do you have soil? Of course, you do. If you have soil, regenerative agriculture will work. That's because the five principles of soil health work anywhere. If you follow the principles diligently and get the soil biology right through your practices, the rest will follow.

As Courtney and I were writing this book, we talked about how to address this question of "Will it work where I live?" And the best answer we came up with was to include the stories of other farmers and ranchers who have tried regenerative methods and experienced their own, sometimes quite amazing, successes. Courtney took the lead on this, interviewing eight farmers and ranchers who changed the way they saw the world after becoming inspired by the possibilities of regenerative agriculture. Although their paths and operations were different, as were their trials and tribulations, each made it work where they live. In this chapter, Courtney tells their stories.

Darin Williams, Eastern Kansas

Darin Williams and his wife, Nancy, are young farmers in eastern Kansas, where they grow a variety of cover and cash crops, including soybeans, corn, wheat, rye, triticale, sunflowers, and a "chaos" garden (like the one Gabe planted on his farm, described in chapter 3). Darin never intended to go into agriculture, even though his grandfather was a farmer and Darin did a bit of farming as a young man. He was told there was no future in farming, so he moved to Kansas City, became a carpenter, and created a successful business as a homebuilder. The desire to farm never went away, however, so in 2006 he moved to the family farm near Waverly and tried to make a go of it on 60 acres. The land was in poor shape and the only way he knew how to farm it was conventionally with full tillage and heavy use of synthetic fertilizer. This method wasn't profitable, but Darin couldn't see any alternative. That's why he kept building houses to earn a living, commuting from the farm.

Things changed in late 2008 when he read an article about Gabe Brown in an agricultural magazine. Darin had begun to dabble in no-till, but even so, his first reaction to the story was "This is baloney." However, when Gabe came to Emporia, Kansas, to speak at a soil-health conference, Darin convinced four neighboring farmers to travel with him to hear Gabe speak. Although all of them were no-tillers to one degree or another, Darin was the only person in the car that day who decided to give regenerative agriculture a try. The rest of them insisted it would never work on their places.

Today, Darin is harvesting 50–60 bushels of soybeans per acre, while some of his neighbors harvest closer to 30 bushels per acre (the county average) using the full conventional model, including all the costs associated with spraying, fertilizing, and GMO seeds.

"For the first three years, my neighbors made fun of me," Darin recalled, "but they don't anymore."

Darin credits his work as a carpenter and builder for teaching him how to look for business opportunities, new avenues for profit, and the value of keeping an open mind. He didn't necessarily need a lot of scientific data to see that regenerative agriculture worked, and he could tell that Gabe was sincere in what he said about his success and wasn't trying to sell the audience something. He liked Gabe's integrity at the Emporia presentation, and that persuaded him to commit to trying regenerative methods for five years. It took only two years to see a positive difference. The soil on his farm had become healthier and his yields were already rising. He knew he was on the right track.

Implementing regenerative agriculture on his farm was a three-step process, Darin said. Step one was to add more diversity. His original operation was solely corn and soybeans, so he knew he had to add a cereal crop in order to diversify the soil biology. He intended to start by planting an eight-way cover crop mix, including turnips, radishes, buckwheat, sunflowers, and millet, in order to add livestock to his operation, but decided to go with oats first, since he knew they had been grown on the farm in the past. A trip to Bismarck to visit Brown's Ranch gave him additional inspiration. Gabe showed him how the roots of the plants in his fields grew straight down, deep into the soil, instead of sideways as they would when they hit compacted soil, typical of many farms.

Once Darin saw the outstanding weed control he accomplished using cover crops, he considered using non-GMO crops. As a test, he decided to follow the cover crops with plots containing both non-GMO and GMO cash crops. This would allow him the best comparison possible. The results were telling. The yields of the non-GMO soybeans matched those of the GMOs. This meant the non-GMOs were returning significantly higher profit due to lower herbicide costs and their ability to bring a significant premium in the marketplace.

Today, the Williamses consistently see yields of their non-GMO soybeans come in at over 50 bushels per acre. In 2017, one farm yielded over 65 bushels per acre without the use of any seed treatment, fertilizer, or fungicide. This is a tremendous yield, especially when one realizes that the local county average is 28 bushels per acre using conventional practices.

Step two was to add livestock. However, Darin didn't own any cattle and wasn't sure what breed to buy. In the end, he settled on British White cattle, and today he direct-markets grassfed beef to customers in Lawrence and other nearby cities.

Step three was to compare the yields of non-GMO soybeans on his farm to those produced using glyphosate. "When I discovered that the yields were nearly the same, that's when the value of healthy soil clicked in my brain," he said.

For Darin, like Gabe, the key is profitability per acre, not yield per acre. Even though cover crop seed costs money, growing covers lowers other input costs, thus saving him money. Not only were the profits higher with the added economic value he was getting, the crops *looked* healthier, which he took as a sign of healthier soil. It happened quickly, too. He saw yields increase every year for the first three years. "Conventional wisdom in no-till says it takes seven years before you will see significant increases in yields," he said. "Not so, as I now know from personal experience."

In 2010, Darin became a full-time farmer, hanging up his builder's tool belt. He still uses some fertilizer and a little bit of herbicide now and then, noting there's no such thing as a "perfect" farm. He's not certified organic, either. "I tell people the only thing I want to be certified at is being successful."

Recently, Darin and Nancy started a non-GMO grain distribution terminal called Natural Ag Solutions, LLC. This distribution center buys and sells non-GMO grains. This distribution process increases the profitability for farmers who grow these grains, which in turn benefits the ecosystem.

Why don't more farmers make the leap like the Williamses have? Darin thinks that part of it is money. Many are in such a tight spot financially they are not willing to take a risk and possibly make a mistake. Part of it is habit. They know how to produce a crop under the current system (with huge help from government crop insurance), so there is not a lot of incentive to try something different. And part of it is a habit of blaming someone else for their mistakes rather than admit that the conventional model of agriculture is working against their long-term viability. Just like his mentor, Darin was willing to take some risks.

"Regenerative agriculture requires commitment and a lot of trial and error to figure out what works on your operation," he said. "You also have to be willing to learn from your mistakes." There's no silver bullet, he insisted, no "one way" to farm, no "secret" cover crop mix that will solve all your problems. "People think you're an expert, but you're not. Nature will always have the last laugh."

The key is to implement the five principles of soil health and be flexible as conditions develop. "Don't limit yourself," Darin said. "Don't box yourself in. I focused on the soil and the organic matter. Everything else is secondary."

Russell Hedrick, Central North Carolina

Russell Hedrick was a twenty-seven-year-old, full-time fireman with nearly ten years of firefighting under his belt when he decided to switch careers and become a farmer. The list of challenges confronting him as he pursued his dream was daunting. As the first person in his family to go into agriculture, he had no prior knowledge or experience to lean on as he took the plunge. Perhaps more importantly, he had no equipment either! What he had was a lifelong interest in farming and a great deal of desire to give it a try.

Russell had his eye on 30 acres near Hickory, located northwest of Charlotte, on which he planned to grow grains and raise livestock. When he asked for advice on how to get started, neighbors told him he needed to buy three pieces of equipment or else he wouldn't be successful: a 150-horsepower tractor, a twenty-foot disc for tilling, and an eight-row planter. Seeking a second opinion, he spoke with Lee Holcomb, a district conservationist with the USDA's Natural

Resources Conservation Services (NRCS) in North Carolina, who recommended no-till and cover crops instead. Lee also suggested Russell speak to Ray Archuleta and Gabe Brown before he made a final decision.

"I was given two completely different perspectives," Russell recalled, "and I didn't know which one to pick, because I'd seen tillage forever and I assumed that's what you did. But a lack of money was an issue, which is why no-till appealed, too."

He called Ray Archuleta first, who worked nearby at the NRCS National Technology Center in Greensboro. Ray explained the benefits of no-till and cover crops in detail and recommended that he call Gabe about livestock. Gabe's first message to him was a basic one: If you're going to farm successfully, you have to learn how to make money at it. The conventional system was all about spending money, he told Russell, not making it. Focus on profitability per acre, he advised. To accomplish this with livestock, Gabe recommended small- to medium-frame animals raised as grassfed and direct-marketed to consumers. As they discussed Gabe's no-till, cover crop, livestock model, Russell wondered aloud if it would work in North Carolina, where the conditions were very different. To his surprise, Gabe said the model would work just fine for Russell. With the fifty inches of precipitation the region received annually, Gabe said, compared to North Dakota's sixteen, it should be easier!

"As for the five principles of soil health," Russell said, "Gabe reassured me that as long as the sun shined on my farm, soil health will work."

Russell decided to give regenerative agriculture a try.

As a first step, he bought cover crop seed and planted it in early October with an eye toward growing corn as his primary cash crop the following April. Fortunately, the land was in decent shape, Russell noted, despite having been in conventional tillage for decades, subjected to the usual assortment of fertilizers and chemicals. Still, the carbon stocks in the soil had been depleted, which is why Ray and Gabe told him to focus on raising the levels of organic matter as quickly as possible. However, when he asked the agricultural depart-ment at a nearby university how much organic matter he could expect to build with no-till and cover crops, their answer came as a shock: *none*. It couldn't be done, they told him, because there is no freeze-thaw cycle in North Carolina (unlike North Dakota), which meant whatever organic matter he managed to add to the soil would simply be consumed by never-dormant microbes. Russell tried to explain to them that an active microbial population in the soil, plus earthworms, meant the carbon cycle would be working all year to *improve* soil health and thus carbon stocks. But they didn't want to hear what he had to say, especially from a newbie farmer.

"To drive home their point," Russell said, "they just told me flat-out that cover crops were a waste of time. Farmland had to be left bare for five months each year to be productive."

Their position recalled advice Gabe had given him: Why pay taxes on land for twelve months but only farm it seven months? Why not grow plants all twelve months instead!

Russell stuck to the regenerative game plan, including the integra-tion of livestock into the crop production. He started with ten cows and twelve pigs, which was all he could afford. Today, he has forty cows, fifty pigs, and a few sheep, too. Yields came up very quickly. In fact, in 2016 Russell won the statewide yield contest for dryland corn at 318 bushels per acre—only four years after learning how to farm! Not only did he easily beat the state average yield at 230 bushels per acre, the winning farmer on irrigated ground that year produced only 2 more bushels per acre than Russell did.

The statistic that he most relishes, however, is this one: Contrary to what the university experts predicted, the soil organic matter on his farm has risen from 2 percent in 2012 to over 5 percent today.

Russell has expanded his farm from 30 to 1,000 acres. In addition to corn and livestock, he grows soybeans, barley, and oats. He has taken Gabe's business advice to heart as well, expanding the variety of enterprises on the farm. Today, part of his annual corn crop is heritage Hopi (blue) corn, which fetches premium prices. Another portion of the corn crop goes to the nearby bourbon distillery (of which he is a co-owner). The density of his individual corn ears, as it turns out, produces more alcohol per volume than conventionally grown corn. Russell is a seed producer as well, selling to multiple markets. Recently, his farm practices have caught the attention of various breweries and malt houses, who have approached Russell about the possibility of growing ingredients for them. To top it off, he is teaching a class at North Carolina State University on regenerative agriculture!

Russell uses social media, both to advertise his products and to share his newfound experience in improving soil health. "We've seen the land come back to life twice as fast with animals than with just cover crops," he said. Russell also admits to being a YouTube junkie. He calculates that he has watched thousands of farming videos since 2012. Because he likes the "free stuff" the internet provides, as he puts it, he uses social media as an educational platform for beginners, as he had been.

"Farmers hear Gabe's story and go buy livestock," he said, "but often they don't know how to do it right. I try to help with social media."

The journey from fireman to farmer has been short and success-ful. Russell credits the integrity and generosity of Ray and Gabe for the farm's productivity and profitability. It is not a coincidence, he firmly believes, that people who care about soil health are also caring people.

"Out of the blue, I called a man named Gabe Brown, who had no idea who I was, and he had no reason to talk to me other than the kindness of his heart."

Jack Stahl, Northwest Alberta, Canada

In 2012, as Gabe gave a talk in Manning, Alberta, Canada, to a group of farmers and ranchers, a Hutterite man sitting in the front row scowled at him the entire time. Gabe has spoken to some tough crowds over the years, but this severe-looking farmer appeared to be his toughest audience yet. However, Gabe was in for a big surprise.

"I was hooked after his first two sentences," Jack Stahl recalled.

Jack had come to the meeting looking for a way to reduce the use of synthetic fertilizer on the large family farm he operated with his brothers near Manning, in northwestern Alberta. The annual cost of fertilizer kept rising while productivity had essentially flat-lined, which meant profit margins on the farm had begun shrinking. They had switched to no-till some years earlier but had to compensate for the increase in weeds with heavy use of herbicides and pesticides, the costs of which were also rising. Jack had begun searching for answers to this dilemma when he heard about Gabe's talk. Jack assumed that Gabe was just another extension agent, and he expected to hear more of the same conventional advice: more fertilizers, more chemicals, more killing things. When Gabe noted, however, that the nitrogen in the atmosphere was free and could be easily employed as a fertilizer for crops, Jack realized he had met a kindred spirit!

"We were about to hit a brick wall economically" he said, "and I knew we had to change the way we did things, so I liked the things he was saying. But listening to Gabe that day I also realized we had to 'unlearn' farming and start over. I know now that successful farming today equals the person who 'unlearns' what they know the quickest."

When Jack returned to the family farm after the talk, his brothers were skeptical but willing to experiment on a "show us" basis. They decided to try a fourteen-way cover crop mix and saw immediate positive results that summer. "Everything came up," Jack said, "and it set my brothers to thinking about a whole new approach." In the next step, they trialed different combinations of cover mixes on different parts of their land, each one successfully. There was no third step—they decided to abandon the experimental phase altogether and "go for it" instead across the entire farm. Before long, their input costs had dropped 60 percent from just a few years before and yields far exceeded expectation.

"This will work anywhere where plants grow and there's soil," Jack said.

At a soil conference in Edmonton in 2015, Jack met with Gabe, Jay Fuhrer, and Nicole Masters, a soil scientist from New Zealand who became the soil analyst on the Stahls' farm. The goal was to make the journey from conventional into regenerative agriculture as short as possible. The farm had weaned itself almost entirely from synthetic fertilizer and had integrated livestock into the crop production. Since Jack had been a fan of both holistic management and ranching for profit for years, the farm had implemented planned grazing principles but had kept the ranch operation separate from the farm component. Not anymore. Today, they use cattle to graze the cover crops as much as possible throughout the year.

Jack acknowledges there is no precise blueprint for managing an agricultural enterprise in this manner but believes that Gabe's trial-and-error model can be condensed considerably, as they are experiencing, thanks to help from Gabe, Jay, Nicole, and others. Yields are up, costs are down, and natural fertility has returned.

"Carbon in the soil is worth way more than money in the bank," Jack said. "I told a banker who was impressed by our crop yields that he ought to be impressed by the carbon levels instead!"

Jack said another important benefit of their new approach is *freedom*. They are free from chemical dependency, greedy corporations, certifying bureaucracies, and heartless marketplaces. They are also free from a destructive mentality that comes with industrial agriculture.

"We cannot keep killing the biology," he said. "We need to restore photosynthesis."

Something else has been restored, too, in the process: fun. "When you change the way you think," Jack declared, "everything becomes fun again."

He loves watching nature perform without the arrogant interference of humans. Grasshoppers are a good example. The insects are a major challenge in his area, but they are no longer a problem on their farm because they can't ingest the high levels of sugar that healthy plants generate by growing in biologically alive soils. This high sugar (Brix) content makes his crops grasshopper-proof. The farm stopped using insecticides years ago and today he doesn't mind at all seeing the critters flying around the land. He considers the use of GMOs to be an act of arrogance, as well. "Over the long run, you can't manipulate nature and win. It will always have the final word."

Even the prolonged drought during the summer of 2017, the worst in modern memory, he said, was fun. Watching how nature responded when the plants and the soil were in a healthy condition was exciting. "Flood, snow, drought, heat—it doesn't matter," he said, "not if Nature is healthy."

If watching your farm endure a major drought might be considered an unusual form of entertainment, Jack admits to being a bit of a contrarian, especially when it involves doing things that people insist won't work. If that sounds like his mentor, Jack will quickly tell you how much he has been inspired by Gabe's ideas and unselfishness and how much obligation he feels to share their success and experience (including an interview, which he normally refuses to do).

Jack says, "I want to be the Gabe Brown of western Canada!"

Jonathan Cobb, Central Texas

Although he represents the fourth generation on his family farm, when Jonathan left for college he intended to never come back home to farm.

He earned a degree in business, got married, and moved to the city. The siren song of farming, however, proved to be too strong. In 2007, he and his wife moved back to his family's 2,500-acre farming operation, located north of Austin. Two generations earlier, the Cobb farm had been a diversified operation, typical of rural America prior to World War II. Over the years, however, it had changed to a modern, industrial farm with monoculture crops and lots of synthetic inputs—not unlike the plight of the farms described in one of Jonathan's favorite books, Wendell Berry's *The Unsettling of America*.

It wasn't all bad news. Jonathan's father used the strip-till method—which is a type of shallow tillage—and had eased up a bit on synthetic fertilizer and biocides over the years in an attempt to save money, which Jonathan said had improved the land somewhat in the decade before he returned to the farm. In fact, carbon levels had risen over ten years from 1 percent to 2 percent, according to soil tests, and water infiltration rates had improved as well.

Except, Jonathan was unhappy. He came home to the farm with fresh eyes and experience in the business world and saw a model of agriculture at work that was financially unsustainable. Yields on their farm, while above the county averages, were not measurably higher than decades prior. Crop insurance claims were being filed nearly every year, either due to drought damage or for high aflatoxins (toxins caused by mold) levels in corn. The land had improved slowly after his father switched to conservation methods, but the long-term economic viability of the farm to support two families was not possible with such high input costs. There was no resilience in the system, he said, which meant it constantly walked a tightrope between success and failure. When a severe drought began in 2011, it pushed the farm—and Jonathan—to a breaking point. He decided it was time to have a talk with his father about calling it quits. It looked like the 110-year old farm had reached the end of its run.

"This is one of the hardest conversations to have in an agricultural family," Jonathan recalled. "I was very emotional."

Fortunately, the farm was saved by an inspired moment of show-and-tell.

In 2011, Jonathan was invited by then–state agronomist William Durham to attend an NRCS training in the region that featured Ray Archuleta as a main speaker. As usual, Ray demonstrated the slake test (described in chapter 3), and the soils from the conventional, high-tillage fields quickly fell apart when they hit the water in the plastic tubes, as they always do. To be successful over the long haul in agriculture, Ray told the audience, your soils must have structural integrity.

Jonathan got it. The dissolving clods and Ray's words made a huge impression on him. "By the lunch break," he said, "I had decided that this was something too important to pass up and that I would stay on the farm."

He had decided to "jump down the rabbit hole," as he put it, of soil health. He introduced himself to Ray at the meeting and later that year met Gabe Brown at a field day on Dave Brandt's regenerative farm in Ohio. Meanwhile, he read every book recommended to him and watched countless videos. He called the experience his "postgraduate education." A passion for soil health developed and farming suddenly became a source of enjoyment again, instead of despair. Fortunately for Jonathan, his dad also went to the NRCS meeting and saw Ray's presentation. This time, the conversation between father and son was different—dad was willing to go down the rabbit hole as well!

They sold the tillage equipment and purchased a no-till seeder, putting in 450 acres of wheat within a month of talking to Ray. Jonathan met Nebraska farmer and businessman Keith Berns, who mentored him on cover crops and how to run a successful business with family. They planted cover crops following the wheat in the summer of 2012, followed by 1,200 more acres that fall. The following year, they brought cattle back on the farm, plus a small flock of chickens. In 2014, they reduced the number of acres in production to make things more manageable under the new system. In 2015, they added hogs and sheep and began direct-marketing grass-finished beef and lamb and pastured pork and eggs. In the meantime, they began to rebuild and restore the native, perennial prairie of the area for their grazing animals, converting it from cropland by employing adaptive, multi-paddock livestock management.

"*Restore* is not the right word," Jonathan observed, "because the prairie was completely obliterated where we are. What we're trying to do is go with the native climax species as well as allowing what thrives in the area naturally today."

Meanwhile, the organic matter content of the farm's soils continued to rise as they implemented regenerative farming practices—today it stands at slightly above 4 percent.

All of this change hasn't been easy. They made lots of mistakes. Pressure was high to "prove that this can work here" in Jonathan's mind. He had taken on this challenge and felt a burden to not fail for the sake of the cause. "In my ignorance, I didn't know what the 'this' was that I was trying to prove could work. I was still thinking inside a broken paradigm of agriculture."

The stress of trying to prove something to others took its toll on Jonathan. Many in his community may have thought his family's "good farm" was going downhill with all this unusual farming. During the darkest of days, the encouragement of fellow regenerative agriculturists carried him through, however. That encouragement along with the continued application of holistic management decision making helped Jonathan and his family focus on reaching goals that were based on a different paradigm.

Regenerative agriculture also helped Jonathan address a spiritual crisis. Back in 2011, before he heard Ray speak, Jonathan had planned to leave the farm to pursue theological studies, another passion of his. He was having a great deal of difficulty reconciling his religious faith with the poor stewardship of God's Earth that he saw all around him. It had been a daily struggle of reconciliation since returning to the farm with his wife in 2007, magnified by hearing a podcast by Dr. Timothy Keller titled "Can Faith Be Green?"

In the end, his spiritual crisis fueled his embrace of regenerative agriculture. His passions returned, bolstered by his family and the strong support he received from the network of soil-health advocates, especially Ray and Gabe. Rains helped, too. Plants came up, new business ventures were launched, and Jonathan became involved with various nonprofit outreach activities, including ones offered by Holistic Management International and the Grassfed Exchange. To remind him of the spiritual goal of their work and that it is a path worth following, albeit difficult at times, Jonathan keeps a quote by

Wendell Berry tacked to a wall in his farm workshop. It reads: "We have the world to live in on the condition that we will take good care of it. And to take good care of it, we have to know it. And to know it, and to be willing to take care of it, we have to love it."

Brian Downing, South-Central North Carolina

Brian had a compaction problem—so he decided to call Gabe Brown.

Brian had wanted to be a farmer from a young age when he "messed around" in his grandfather's garden, sold vegetables at his school, and cared for a few cows that his family owned. He studied to become a pharmacist in college but quickly decided against it as a career. He pursued an animal science degree, which landed him at an 1,800-head Holstein dairy for nine years and taught him a great deal about the pros and cons of large-scale food production.

When his grandfather's small farm became available three years ago, Brian purchased the property and buckled down to become a farmer—until he realized he had a problem.

Located on a slope near a river with heavy clay soils, the farm had been conventionally no-tilled and overgrazed by livestock for forty years, resulting in compacted soils with poor water infiltration. As a consequence, crops suffered from a lack of moisture, though the lack of soil structure also meant the land was prone to flooding during heavy rainfall events. Making matters worse, many North Carolina farms have been degraded from monoculture tobacco production, which has resulted in an excessive buildup of phosphorus in the soil. Leery of conventional solutions to these problems, Brian decided to call Gabe, whom he had seen on YouTube videos.

"To my great surprise," he said, "not only did Gabe answer the phone, but he spent an hour talking to me about the farm!"

Together with Ray Archuleta, who worked nearby Brian's farm at the time, they came up with a low-cost plan: Brian planted a five-way cover crop blend that fall, including cool-season grasses, cool-season broadleaves, and brassicas, and then grazed this cover crop twice with cattle using high stock density before terminating it in the spring with a homemade roller-crimper. It worked! Animal integration not only provided a much-needed ecological kickstart, Brian said, but also provided low-cost grazing for his cattle as well. Striving for ecological recovery on a budget provided immeasurable benefits, validating Brian's choices.

In the next step of the plan, he managed the small herd of cattle on the cropland according to holistic, planned grazing principles in order to further stimulate perennial plant growth. He saw positive effects almost immediately. Additional cover crops established themselves nicely, smothering the weeds that had dominated the farm previously. Infiltration rates rose, and dense clumps of earthworms appeared, followed by rapidly disappearing crop residue on the surface.

"Those suckers got hungry," Brian said.

The soil was softer, he noticed, and spongier. The organic matter rose 2 percent in three years in fields where the livestock were integrated. He decreased his fertilizer use by 50 percent and eliminated all pesticides and fungicides. His use of herbicides has dropped to once a year.

"If I do my job right, I don't have to worry about weeds," he said. "If I do have a weed, I'm not worried about it. It's there to remind me of my mismanagement, that I didn't follow nature's principles."

Under the plan, production on the farm has both increased and diversified. The same 40 acres that was once grazed by only twenty brood cows now supports a menagerie of livestock: thirty-five steers for grass-finishing, three hundred laying hens, pastured broilers, pastured pork, and 8–10 acres of vegetable production.

"This level of production would never be possible with conventional practices," Brian said. "Diversification not only helps ecology, it helps risk management as well."

At first, Brian didn't move the cattle as quickly around the farm as Gabe had suggested, though he's doing that now. He is also focused on producing nutrient-dense food, which his customers appreciate. One customer texted him shortly after buying eggs from the farm for the first time. She wanted more. "I've never had eggs like that before in my life," she wrote.

It's all added up to a strong feeling of satisfaction. Brian admits to be being obsessed with a desire to heal the planet, a desire compounded by the knowledge that his farm is surrounded by new residential homes. Regenerative agriculture is the proper way to accomplish this necessary healing, he sees now. It's not just wishful thinking. Brian has a day job teaching agriculture to sixth-thru-eighth graders, an educational program available at every middle school in Randolph County. His enthusiasm for regenerative agriculture has led him to attempt to rewrite the North Carolina middle school agricultural curriculum. Brian has also approached the national Future Farmers of America organization to see if they will consider removing the plow symbol from their logo!

"Einstein warned us against doing the same thing over and over while expecting a different result," he said. "And when I look into the future, what I see is a world that desperately needs a different result."

Axten Farms,
Southern Saskatchewan, Canada

When Gabe was a supervisor on the Burleigh County Soil Conservation District, he liked to attend the annual Soil Health Tour, a drive-it-yourself tour of farms in the county that were working to advance soil health. While standing in line for the evening meal on the tour several years ago, Gabe overheard two men in front of him talking about wanting to learn more about soil health. Not being shy, Gabe interrupted them and introduced himself. They explained that they were on their way to visit Dr. Dwayne Beck at the Dakota Lakes Research Farm in South Dakota the next day and they decided to take in the tour. Gabe visited with them over supper and invited them to stop at his ranch on their way back home. One of those fellows was Derek Axten.

Derek and Tannis Axten, along with their children, Kate and Brock, operate Axten Farms near Minton, Saskatchewan. They grow 5,500 acres of grain crops in a part of southern Canada that receives only twelve to fifteen inches of precipitation annually. No-till was a common practice in the area, but improving soil health was not, which meant they used a large amount of synthetic fertilizers, pesticides, and fungicides. Derek said he took the trip because they could not continue to farm with such high inputs and low margins. Everything he saw on that trip resonated with him. And on his return, after he explained what he had learned to Tannis, she was on board, too. It helped that Tannis was a biology teacher and had a good understanding of the importance of microbiology!

As they will tell you, their journey since then has been amazing.

First, they learned as much as they could about how soil functions and the importance of biology in soil health. They attended classes taught by Dr. Elaine Ingham, one of the pioneer researchers and educators in the importance of the soil food web, and Dr. Wendy Taheri, a leading soil scientist. The Axtens recognized the importance of taking the time, and money, to invest in education. Tannis oversees the monitoring of the soil biology on the farm and their compost program. She pays special attention to mycorrhizal fungi, and one step they took right away was to discontinue insecticide use and seed treatments.

Derek set out to experiment with different cash crop rotations, cover crop mixes, and intercropping strategies. (Intercropping is the practice of growing different crops together.) Durum wheat and lentils are the predominant crops in their area, but Derek knew that he needed to diversify. He added oats as a cereal crop and mustard, red lentils, forage peas, mustard, flax, and chickpea as intercrops. Giant green lentils, faba bean, and fenugreek were seeded as companion crops. The practice of intercropping is becoming more common as producers recognize the advantages of both crops to soil health and profitability.

To add carbon to the system, cover crops are integrated into the rotation. Derek converts these covers to dollars by custom-grazing cattle on them during the fall and winter months. He uses a diverse cover crop blend of cereal rye, teff grass, daikon radish, turnips, chickling vetch, flax, and red clover. This feeds the soil biology a diverse diet and advances soil health.

The Axtens pay close attention to the amount of armor (residue) on the soil surface. In their dry environment, armor is critical not only to protect the soil but to keep evaporation rates and soil temperatures down during the heat of the summer. It also inhibits weed growth and protects against erosion.

What is the result of all of these changes? Derek said input costs have been substantially reduced, leading to increased profitability per acre. "Farming is fun, and we have more time to spend with our children," Derek explained.

To top things off, Derek and Tannis were named Saskatchewan's Outstanding Young Farmers for 2017, a true testament to the importance of focusing on soil health.

Joe and Ryan Bruski, Southeastern Montana

Anyone who thinks regenerative agriculture principles cannot work in a dry environment should talk to the father and son partnership of Joe and Ryan Bruski. Joe has farmed and ranched all his life in the sandy, semiarid environment of southeastern Montana. For years, he made a living growing small grains and running a herd of beef cattle while always aware that they were only a few weeks away from a drought. Few places in the nation are as challenged by the inconsistent nature of the precipitation as Ekalaka, Montana.

Joe's son, Ryan, went off to study farm and ranch management at Bismarck State College, where one of his instructors was Gabe's son, Paul Brown. As part of the course, Paul brought his students out to

Brown's Ranch, so they could see the soils, plants, and animals for themselves. Ryan took an interest, especially in the soil health aspect of the course. Paul could tell that Ryan was eager to learn more, so he asked Ryan if he would be interested in working on the ranch after school hours. This gave Ryan the opportunity to learn more about the soil-building principles that Paul taught.

Ryan was not only a good worker, but he was an observant student as well. He quickly learned the five principles of soil health and shared his knowledge with his father. To Joe's credit, he listened to his son and gave Ryan the opportunity to apply those principles on their own operation.

Joe had been spending over $100,000 a year on synthetic fertilizer for their 3,500 acres of cash and hay crops. He knew, however, that they could not continue to operate that way, given the low selling prices of commodity crops. There just was not enough profit potential to justify those inputs.

Another problem was that for decades, their crop rotation had lacked diversity. Many acres were hayed each year, with no biomass returned to the soil. As Ryan said, "This was turning our sandy soils into a desert." He knew that he had to increase the diversity, biomass, and amount of liquid carbon available in order to feed the soil biology. With this in mind, they planted a diverse cover crop mix with the intention of having their cows graze it during the winter. For years they had spent their summers making hay and their winters feeding that hay to their cows. Having a cover crop for the cows to graze for at least part of the winter was a big, positive change. It provided diversity, and with proper grazing management, they were able to leave armor (residue) on the soil surface, which was critical in their semiarid environment.

Their new grazing management not only saved money, but it improved the quality of their lives—they had more free time! Their next decision was to increase the number of acres seeded to covers. It didn't take long for them to see results. Despite dry conditions, the organic matter level improved. From a starting point of 1.7 percent in 2008, it increased over a tenth of a percent each year, which is a very good increase for sandy soils. They also increased the size of the cowherd. Having winter grazing for the animals significantly lowered feed costs, too, all while advancing the health of their soils.

On the Bruski ranch today, the no-till drill is hooked up in the spring and stays hooked up until the snow flies. Cool-season cover crops such as peas and oats are seeded early. Diverse warm-season species such as sorghum/sudangrass, millet, and cowpeas, along with a forage brassica are seeded in the summer. Every time it rains, more cover crop is seeded, thus ensuring the presence of living roots in the soil as long as possible throughout the year. At times, however, not enough rain falls to provide for adequate growth. Ryan is quick to point out that they do not consider this a failure. Even though the plants may not be tall enough to graze, they are providing a valuable service, feeding soil biology and protecting the soil from wind erosion and moisture evaporation. In the fall, they seed biennials such as winter triticale, forage winter wheat, and hairy vetch. This crop enables them to take advantage of the moisture that is received in the form of snow, and it can finish its growing cycle before the summer heat occurs.

As soil health advanced, so did the ranch's ability to withstand weather extremes. This was evident in the very dry conditions that persisted in the 2017 growing season. While most neighbors were haying spring wheat crops that yielded less than 5 *bushels* per acre, the Bruskis fall-seeded winter triticale crop yielded an astounding 3 *bales* per acre. This was a testament to the Bruskis' focus on soil health. Their neighbors took notice!

"By following the five principles of a healthy soil ecosystem," Ryan said, "we have made our ranch much more resilient to drought. And in our environment, the question is not if a drought is going to come, it is how soon."

They have noticed an improvement to their perennial pastures as well. Having cover crops to graze allows more recovery time for their pastures, thus leading to stronger, healthier plants that can now withstand drought. Not coincidently, "native" species are returning in both diversity and quantity.

The Bruskis are increasing diversity not only of plant species, but of livestock as well. Besides the increase in the cowherd from four hundred to eight hundred head, they added a yearling (stocker) operation and Ryan has built a reputation of selling high-quality pastured pork. The additions of laying hens and goats have also expanded their diversity. All of these add income streams, thus improving both cash flow and resiliency.

Perhaps the most important benefit of the Bruskis' move into regenerative agriculture, however, has been the improvement in their quality of life.

"It has made farming and ranching fun again," stated Joe. "I now have time to do things I want to do, including spending time with my granddaughter!"

Gail Fuller, East Central Kansas

Rancher Gail Fuller is a believer in the famous quote from Franklin Delano Roosevelt: "The Nation that destroys its soil destroys itself." And Gail's biggest pet peeve about farming is erosion. To combat this persistent problem, he tried no-till in the 1980s but couldn't make it work. He returned to conventional tillage with the exception of one 10-acre field. In the early 1990s that no-till field was still producing well—and not eroding. Gail began to rethink his decision to till.

A spring flood hit the farm in 1993, and after the water receded Gail was shocked and dismayed. One field that had been tilled just before the flood had lost eight inches of soil! There was an obvious drop in the soil level at the spot where the tillage had stopped. After seeing that, Gail parked his tillage equipment for good.

While no-till was becoming a little more mainstream in Kansas in the early 1990s, it was still a difficult management skill to master, at least for Gail. By 2002 he was ready to give up on no-till again because erosion was still occurring on his farm and he was struggling to achieve decent yields. At the advice of a friend, and against his better judgment, he planted one field to wheat that fall rather than leaving it fallow. The wheat cycled carbon. It was an "aha" moment! It helped Gail to realize that no-till was just a tool. It was a piece of a larger puzzle. He suddenly understood that the old system (corn and soybean, with the corn chopped for silage) was starving the system of carbon. He had tried growing cover crops in the late 1990s, but he hadn't understood their importance back then, and when drought hit the farm in 2000 and money was tight, the cover crops were the first to go.

From that day forward Gail's emphasis was on how to get as much carbon into the soil as possible. He brought cover crops back onto the farm. He also brought cows back onto the land (they had been kicked off when he started no-till because he had been told no-till equals no livestock).

Wheat was Gail's most profitable crop, but in 2007 he tried planting an oat/pea cover crop mix in late February. He chopped the mix for silage (bad habits die hard) and then planted corn on that field in late May. (That was six weeks behind the normal planting date in eastern Kansas.) Gail also cut back on nitrogen, applying a little more than half the normal amount. When the corn yielded 199 bushels per acre, Gail knew he was onto something. He repeated this crop sequence the following year, and for seven years it worked well, providing his highest yielding corn.

On a 2010 field day on his farm, Paul Jasa from the University of Nebraska at Lincoln had set up a rainfall simulator using Gail's soil. The results were far from spectacular. "Where's all your residue, Gail?" asked Paul. From that day forward, Gail stopped chopping cover crops for silage. Instead, all covers were grazed or left in place to feed soil biology.

The drastic changes Gail made on his farm were putting him in the spotlight locally, and many of his neighbors were skeptical. They would call him in February to ask what he was planting. When he told them he was sowing oats and peas to feed the soil, they didn't understand. Even after his methods started to prove successful, no one wanted to follow his lead. Gail has long said, "Peer pressure in agriculture is our biggest hurdle to converting to regenerative agriculture." How did Gail manage the peer pressure?

In 2005, he read an article about a farmer who was doing the same crazy things as he was doing. That farmer was Gabe Brown. That winter Gabe came to Kansas to speak at the No-Till on the Plains Conference, and Gail was sitting in the front row. "I followed him everywhere," Gail said. "I was starved for information." This started Gail down a path of relearning how to farm. Each thing he learned brought the picture a little more in focus.

As Gail's soil began to heal he noticed that his crops were looking healthier, too. One evening at a conference, he, Gabe, Jill Clapperton, and others discussed what this meant in the big picture. If our crops look healthier, and the soil is healthier, would the grain not be nutritious? This was a new idea to Gail, and an exciting one.

Gail's girlfriend, Lynnette Miller, had been doing research of her own. She was concerned about the growing health crisis in the United States, and was also sick of eating bland store-bought eggs. Soon Lynnette had chickens roaming the farm, and next she wanted to raise some sheep. Gail relented.

As Gail began to dig deeper into soil health and nutrient density, Lynnette was digging into human health. Gail had begun to follow the work of Dr. Don Huber, but he wasn't sure he believed all the alarming claims Dr. Huber made about the harmful effects of glyphosate. Gail kept thinking, *There is no way it can be this bad. The regulators wouldn't allow it, would they?* But one night he was listening to a radio interview Dr. Huber had done a year or so earlier. It was the third time Gail was listening to the interview; something Dr. Huber said was very important, he knew, but he couldn't put his finger on it. Then, it hit him like a ton of bricks. Glyphosate is an *antibiotic*! That one statement galvanized Gail's thinking. Continually spraying your soil—and thus your food—with an antibiotic could not be good for your soil, or your gut!

Gail had another revelation when, in 2012, he met Dr. Jonathan Lundgren at the No-Till on the Plains conference in Salina, Kansas. Gail soon came to understand the critical role that insects play, not just on his farm, but on the planet. As his understanding of the relationships among soil, insects, and microbes grew, Gail began to focus on the whole: All management decisions should be made with the *ecosystem* in mind. What effects would a decision have short term, and long term, on the ecosystem? "Every ecosystem on my farm is reliant on all the species within it, save one," Gail realized. "And that species is us. If we don't learn to live within an ecosystem, it is we (humans) that are in danger."

Today Gail and Lynnette are on a journey to reverse the ecological damage done on Gail's farm, and to do their part to reverse the damage that has been done to our collective health. The fact that our children's life expectancy is decreasing is a weight on his shoulders, and he is determined that further degradation of soil and human health is not going to happen on his watch.

"Regardless of the question," Gail says, "soil is the answer!"

Ten

Profit, Not Yield

I practiced the conventional production model of farming for many years. I chased higher yield when growing crops and more pounds when raising beef. Everywhere I turned, the message of increasing production was pounded into me. Magazines, newspapers, radio, universities, extension service, agricultural agencies, everywhere and everyone was telling me that I had to produce more "to feed the world." Stacked GMO traits, hybrid grain varieties, foliar fertilizer, seed treatment, larger equipment. As I write this, I am watching my neighbors pull into a field with three large combines, two grain carts, and four semi tractor-trailers. My in-laws farmed for thirty-five years and never had any equipment larger than two single-axle grain trucks, the largest had a sixteen-foot box. My, how things have changed.

It is the same with livestock: performance-tested bulls with the highest expected progeny differences (EPDs), genome testing, total mixed rations with the latest ionophores, all designed to produce more, more, more! I distinctly remember flagging down a neighbor on the road once to have him follow me back to my corrals so I could proudly show him a bull calf that weaned at over 900 pounds. I was so proud!

I chased that model for over twenty years. However, I began to question myself after the disaster years. Those four years of drought and hail were hell to go through, but they were the best thing that could have happened to me. Adversity forced me to change the way I looked at things. Slowly, over time, I found myself questioning why I was following the more-more-more mantra. Was I chasing short-term gain at the expense of my ecological resources?

My doubts and dissatisfaction came to a head one day in 2010 when I was lamenting to Paul the fact that our corn yields, although good, were not as high as some others in the area. Paul looked at me and said, "Dad, don't you think that you are trying to outpro-duce our environment?" WHAM! That statement hit me like a ton of bricks. He was absolutely right. Nature does not care about yield and pounds, nature cares about enduring. Nature wants to be sustainable. Did I want to farm for one more year or did I want to farm for decades? I had to let go of the yield-and-pounds mentality.

Farming Against Nature

The changes in agricultural land in the United States due to the current production model are disturbing and sad. I will use my ranch as an example. From historical archives we know that 140 years ago this part of North Dakota was covered with a diverse mix of cool- and warm-season grasses and broadleaved plants. European immigrants moved onto these prairies, bringing with them the plow. Diverse prairies soon were turned under with tillage. As described in chapter 7, tillage crushes, smashes, and pulverizes soil aggregates.

The tillage continued over many decades, and along with it came the widespread practice of monoculture grain production. Not just monocultures but fewer and fewer crop species as well. Where once grew more than one hundred species, now only a few grow. Overall, just fifteen crops supply approximately 90 percent of the plant-based foods we eat today! The early settlers ate a much more diverse diet than we do.

We can see the loss of species diversity in our commercial vegetable production, too, where we lost well over 90 percent of our vegetable seed varieties during the twentieth century. In 1900, there were nearly 550 varieties of cabbage available in the United States; today only 28 varieties are sold commercially. With beets, the change is from 288 varieties to 17. For cauliflower, from more than 150 to only 9. And corn? I hate to tell you, but we have lost over 96 percent of the corn varieties available at the start of the twentieth century.

Soil scientist Dr. Wendy Taheri has recently discovered that many of today's "new and improved" grain varieties do not have the ability to form symbiotic relationships with mycorrhizal fungi. These varieties are not able to take advantage of all the benefits that fungi have to offer. Breeders have been selecting for traits such as yield and not noticing that in the process other traits—such as the ability to form relationships with fungi—are lost. It's not that surprising, because plant breeders develop and propagate new varieties in sterile soil. The roots have never been exposed to mycorrhizal fungi, thus it goes unnoticed when a variety does not develop the ability to interact with fungi. Those varieties will be fully reliant on applied synthetic nutrients!

Loss of biodiversity has led to less nutrient cycling, which also equated to an increase in the use of synthetic fertilizer. This led to an increase in weeds (most weeds are high nitrogen users). An increase in weeds led to an increase in the use of herbicides. Many of the herbicides used today are chelators. Chelators bind to metal. Metals such as zinc, manganese, magnesium, iron, and copper. Can you guess where this is leading?

These are the same nutrients that plants need in order to ward off disease. A lack of these nutrients can lead to a higher incidence of fungal diseases. An increase in fungal diseases leads to an increased use of fungicides. Fungicides are detrimental to soil biology and pollinators. Yes, pollinators! Recent studies show that fungicides, once thought to have no ill effects on bees, do have an impact. Scientists and corporate executives have to acknowledge that these compounds are having greater harmful effect than we realized. Farmers need to be educated on better methods, and consumers have to demand that the use of these fungicides is discontinued.

The lack of nutrients available to the plant also makes the plant more susceptible to pests. An increase in pest pressure leads to an increased use of pesticides. Of course, the majority of pesticides are not pest specific, which means many beneficial insects will be killed also, including pollinator species, such as bees, that are needed to pollinate our crops. Almost all fruits and vegetables grown on conventional farms today are sprayed with copious amounts of insecticides. Is it any wonder that we have such a dysfunctional ecosystem?

From a livestock perspective, the goal of producing more and more pounds per animal led to raising animals in confinement. Dairy cows and beef cattle were taken off pasture, where they once benefited the ecosystem by grazing living plants, thus cycling more carbon. Instead, today they are raised in confined quarters. Their diets were changed by the all-knowing authorities from forages to high-starch grains, affecting the animals' health and longevity. Most dairy cows raised in confined high-production systems have a life span of less than four years! And the milk, cheeses, and other dairy products from this system are much less nutrient-dense, which is impacting human health as well.

The high starch rations fed to beef cattle in feedlots negatively impacts the life of the animals, too, as well as the nutritional value of the beef itself. Take the example of omega fatty acids. Research has shown that foods with a lower ratio of omega-6 to omega-3 fatty acids are better for human health, and research has also shown that grassfed beef has that low ratio, whereas grain-fed beef has a much higher ratio of omega-6 to omega-3 fatty acids.

The feedlot industry has nothing to do with the cattle business. Feedlot operations are in business to market feed and pen space. They want cattle that will eat a lot of feed and take a long time to finish; that's what is in their best financial interest. Does anyone truly believe that grazing animals prefer to be in a feedlot? Just open the gate and see what the animals choose.

The industry moved hogs, chickens, and turkeys into buildings in the ruse that the animals would be "better off." Did anyone ask the animals? Back in 1983, when Shelly and I first moved to the farm, I took a part-time job at a nearby egg operation. My job duties included cleaning dead hens out of the cages. I started every morning at 6 a.m., kneeling on a trolley, pulling myself alongside the rows of cages, which housed over twenty thousand hens, elevated above tons of fecal material. Nine hens were crammed in a three-foot by three-foot cage, living their lives in an area where they could hardly turn around, never seeing or feeling the outdoors. I wondered about how they must feel, never having the opportunity to scratch through the leaves or catch a grasshopper. They had no opportunity to be a chicken! Right then and there I vowed never to have chickens, at least not chickens in confinement.

The US government has propagated this mindset with its cheap food policy. It wants to ensure that citizens have an abundant supply of cheap food. Notice I did not say *nutrient-dense* food. The United States spends more on health care than any other country in the world, and yet, its citizens are not healthy.

Are farmers and ranchers to blame for all this? No, not entirely, but we need to take our fair share of the blame. The American public needs to take their part of the blame, as well, for allowing this to happen. Through their buying dollars, consumers have made the choice that they want this system, even as they choose to ignore the environmental degradation, the mistreatment of animals, and the overall decline in human health.

And think of what else this production model has caused. It has led to tighter and tighter margins for producers. Lower margins mean producers must farm more and more land to make ends meet. Farm sizes increase, leaving fewer farms overall and fewer people operating the land. In other words, this production model has also led to the demise of many of our small towns. Consider these facts:

- Three companies control over 75 percent of the agrochemical industry.
- Three companies supply over 90 percent of the breeding stock of layers, broilers, turkeys, and pigs.
- Four firms control between half to three-quarters of all animal slaughter, depending on species.
- Five companies control over 50 percent of the farm machinery market.

Paul Aackley, longtime friend and regenerative Iowa farmer, summarized the ramifications of the current production model. Paul writes, "From memory and records since 1949: From where I sit at this computer, there were four occupied farmsteads along the mile of road to the north and a school house at the end of that mile, to the south within half a mile was the school house where I and four others started kindergarten in the fall of 1950 (one later became head of neurosurgery at Yale) and an occupied farmstead within the next half mile and three-quarter mile east and west. Even tenant farmers had a connection to the land, not as strong as owner opera-tors, but a connection. I don't think anyone had figured out how to drain our side hill seeps (wet area) with tile. That came during the decade, but most didn't think they could afford the cost for another fifteen to twenty years. The local weekly paper always had at least one farm sale during the winter and early spring months from the mid 1950s until the mid-60s. There was cost share assistance to apply lime and seed alfalfa/grass in the early '50s. This all began to change during the '60s. NPK became readily available. Atrazine for corn and Amiben for soybeans were introduced. Out-of-county or out-of-state investors began to purchase land that had been marginal, tile out wet areas and clear trees and crop-share or cash-rent it for row crops. Profit from the rent or profit from the inflating land value drove the change. I remember one day watching a 24-row planter work on some adjoining land and realized what the American Indian must have felt when the white man showed up. The land ethic gathered dust on a shelf somewhere. Change (technology) came faster and with more force (profit) than humans could handle intelligently."

Well said, Paul.

The Trouble with Subsidies

The federal crop insurance program was instituted on February 16, 1938. As with most government programs, good intentions often lead to disappointing results. A program that was intended to minimize risk has become a monster that now dictates most of the cropping decisions made in the United States today. In my opinion, this program also ensured that low commodity prices would be the norm for decades to come.

I contend that over 95 percent of planting decisions farmers make today are based on how much money they can guarantee themselves by insuring through crop insurance programs. Farmers know exactly the minimum amount of gross dollars per acre they will receive that year from crop insurance. Keep your expenses below that amount and you will make a profit. What other business is offered those guarantees? Certainly not Ma and Pa's restaurant on Main Street!

This also drives input suppliers to charge more for their products because they know that farmers are guaranteed this revenue stream. Thus, the suppliers charge what they know farmers can pay in order to ensure their own healthy profitability. Fertilizer, herbicides, pesticides, fungicides, equipment, and the list goes on, all continue to increase in price. I can always tell when a local fertilizer or chemical dealer hires a new salesman, because I get a visit from the newbie. I notice that many of these folks drive brand new pickup trucks. Where does all the money come from to buy all of those new pickups?

By the way, one expense I now have that I did not have when I farmed conventionally is that I have to buy my own caps! It seems the salesmen do not give caps to people who do not buy their products.

The profit offered by revenue insurance is quickly gobbled up by industry. Tighter and tighter margins mean that producers are more reliant on the subsidies provided by government programs—programs such as crop insurance, the Environmental Quality Incentives Program (EQIP), the Conservation Security Program (CSP), and a myriad of other acronyms. I took advantage of these programs for many years. I received cost share for tree plantings, fences, wells, even auto-steer for my tractor! I didn't give it much thought, early on anyway. But as I started thinking holistically, as I began to realize the ramifications of taking advantage of these programs, I began to have serious reservations about accepting that cost share.

Ma and Pa's restaurant on Main Street was not getting their insurance premium subsidized. My relatives who live in town had to purchase trees from the nursery at full price; they were not eligible for cost share. What entitled me to receive this money? Again, I was told it was because I was "feeding the world." But the more I thought about it, the more it seemed to me that these payments were just a form of welfare, and I did not want welfare! Shelly, Paul, and I made the decision that we would no longer accept any agricultural payments. No crop insurance subsidies, no EQIP, no CSP. Period.

I found that decision very liberating. I no longer have to spend time at an NRCS or FSA office filling out forms, and more important, I am free to make cropping and other management decisions that are in the best interest of my ranch and my family. I am no longer "tied" to decisions based on what someone else thinks is best. Don't get me wrong: I believe a case can be made for some government programs, especially ones geared to helping young producers or our military service veterans enter regenerative agriculture. I just believe it is inherently wrong to routinely use hard-earned taxpayer dollars to subsidize agriculture or any other business for that matter.

A good case in point was the summer of 2017 here in North Dakota. We did not receive our usual spring rains and the weather turned hot. In early June we had several days in a row over 100°F (38°C), and then, less than forty-eight hours later, it froze! This combination took its toll on forage production. Many producers scrambled to find enough forage for their animals and ended up traveling hundreds of miles to find grazing land or to buy hay. Thousands of cattle were sold off. One of my neighbors had his cattle out on pasture less than two months before he had to start giving them supplemental feed. His pastures looked like a well-groomed golf course! A disaster declaration was made, and the government started the money train flowing.

Producers were encouraged to sign up for this disaster assistance. I was asked to do likewise. Why, I asked? Due to our adoption of Holistic Planned Grazing, we saw few ill effects. We grazed just as many animals as we had the previous year. I adapted my grazing strategy accordingly. I stopped moving the cattle every day, thus allowing them to eat more of the forage. This ensured that I would have forage to move them to. The fact that our pastures had strong root systems and plenty of armor covering the soil helped tremendously. The years spent focusing on creating a healthy ecosystem paid dividends. It had made my pastures resilient to these one-year swings in precipitation and temperature.

I could have collected tens of thousands of dollars simply by walking into the FSA office and signing my name on a form. But I just could not morally or ethically do that. We simply did not have a disaster on our operation. I was not going to fleece the American taxpayers out of their hard-earned dollars just because the system said I could.

In this case I see what happened in North Dakota as less a natural disaster and more a human-caused disaster resulting from poor management. If our agricultural system had not caused a huge drop in soil organic matter over time, our farms and ranches would be much more resilient in times when rainfall does not come through.

As I look back and think of all the practices I received subsidies for in the past, I can honestly say that if I had it to do over again, I would not make the same choices the same way again. To state it simply, I should not have accepted those cost share dollars. Let me give you an example. We received a lot of cost share dollars for fencing. We built over one hundred paddocks on our ranch using permanent fencing. We are now tearing out nearly all of those fences! They limit our ability to regenerate the soil with Holistic Planned Grazing. Too many acres are either under or alongside the fence, and we are not able to get the hoof impact alongside and under the permanent fences. Because of the permanent fences, soil health and forage production suffer. If we were to use temporary electric fence instead, we could move the fence to a different location each year, thus allowing the animal impact soil needs on every square foot of our pastures. We should never have put up all of that expensive permanent electric fence. It was a waste of my time and a waste of taxpayer dollars!

Agencies such as NRCS are trying to do the right thing, but it would be a better use of money and staff time if they would educate producers about how ecosystems function. They should be providing the technical expertise that is sorely needed, instead of overseeing programs that end up being nothing more than entitlements. How much of our federal debt crisis could be avoided if the government phased out such ineffective programs? I know many NRCS employees who are excellent teachers and who could make a much greater impact educating farmers and ranchers, which, in turn, would lead to a much greater financial reward for the farmer or rancher than do those cost share practices. Those employees would much rather devote their time and talents to educating growers than to overseeing programs.

Side-Stepping the Underlying Problems

Production agriculture as it is practiced today is all about applying Band-aids to a larger problem. Water quality is an example. It is common throughout much of the Corn Belt to install tile drainage. A tremendous amount of time, dollars, and resources are spent to manage what producers view as "excess" water. But the first question to address is this: Why is there excess water on their land? Could it be because the soils have been degraded, and thus water cannot infiltrate the land? Could it be that the farmers are not planting a diverse crop mix and have low cropping intensity, and that there are not enough living roots in the soil for more than a third of the year to take advantage of the moisture?

Think about what happens when "excess" water is moved through a tile drain and into the watershed. Most likely, that water carries a large quantity of both nutrients and chemical residues with it. Do these then have an impact on water quality downstream? What effect does this have on fish, wildlife, and people? We, as producers, need to realize that every action we take on our operations has compounding and cascading effects. Could it be that we are negatively affecting the health of our children and grandchildren?

Take a look at these statistics: Americans are at the top, or near the top, in incidence of chronic diseases such as attention-deficit disorder, Alzheimer's disease, cancer, osteoporosis, obesity, autoimmune diseases, and others. However, even when people "eat right," they still suffer the consequences of today's industrial farming practices. As Michael Pollan states in *An Omnivore's Dilemma*, much of what we eat today is not food, it is "edible food-like substances."

In this case, you are what you *don't* eat. Today, many foods and food products on offer at the grocery store are sadly lacking or unbalanced in essential nutrients, including proteins, vitamins, minerals and plant secondary metabolites. And if the foods we eat are deficient in nutrients, then our bodies will end up nutrient-deficient, too.

Studies have chronicled the steady, decades-long decline of dietary minerals in vegetables, including copper (down 24–75 percent), calcium (down 46 percent), iron (down 27–50 percent), magnesium (down 10–24 percent), and potassium (down 16 percent). Potatoes have lost 50 percent of their copper and iron content in the past fifty years, and carrots have lost 75 percent of their magnesium. There have been sharp declines in other nutrients as well, including protein, riboflavin, and vitamin C. Another study noted that you would have to eat eight oranges today to get an equivalent amount of vitamins that your grandparents would have enjoyed from a single orange in their youth. It's the same for meat—you would need to eat nearly twice as much beef, chicken, or pork to acquire the same levels of certain nutrients you would have two generations ago. Nutrient depletion can be traced back to the origin of agriculture ten thousand years ago when the first farmers began manipulating the starch and sugar content of early crops to make them sweeter and less difficult to chew than their wild ancestors, inadvertently reducing key nutrients as a result. Modern, industrialized agriculture has sped up this process dramatically.

Nutrient depletion has become a global crisis. One-third to one-half of the global population is believed to be chronically deficient in essential minerals. Adequate amounts of iron, for example, are essential to ward off anemia, a common blood disorder, but more than one billion people around the world are currently suffering from a lack of trace iron. As I have pointed out, virtually all of the world's agricultural land exists in a degraded condition, which reduces the availability of nutrients in the soil. In Africa, it is estimated that forty million people are trying to survive on formerly fertile land that has degraded to the point where it is essentially nonproductive. In 2006 the United Nations created a new category called "type B malnutrition," which is defined as a diet adequate in terms of the amount of calories and protein consumed by an individual but inadequate in terms of essential minerals and other nutrients. And this is the diet characteristic mostly of industrialized nations, such as the United States! More than two-thirds of American adults and nearly one-third of children are overweight or obese, abetted by the food industry's peddling of calorie-dense diets that depend heavily on high-fructose corn syrup and fatty soybean oil, two staples of industrial agriculture. We are overfed and undernourished.

Many Americans struggle daily to get a sufficient quantity of diverse nutrients into their bodies. The reasons for this are complex, ranging from poor dietary choices by consumers to the narrowing of our food options, which are a result of the current production model. There is also a bias in plant selection by industrial agriculture that favors appearance, growth rate, ease of transport, pest resistance, and shelf-life over nutrient content.

In 2002, the *Journal of the American Medical Association* concluded that diet alone, sourced from the conventional production model, could no longer supply adequate amounts of nutrients and advised all adults to take one multivitamin per day, reversing a long-standing position. The sales of supplements have since grown into a $30 billion-a-year industry. According to the US Bureau of Labor Statistics, in 2014 there were sixty six thousand professional nutritionists at work in the nation, with a projected increase of eleven thousand jobs over the next decade. Estimates for the annual cost of diet-related illnesses (direct and indirect) in the United States range from $250 billion to *$1 trillion.*

As I have mentioned, we host a lot of visitors at our ranch, many from overseas. I often ask the foreign visitors what differences there are between their home country and the United States. Hands down, the comment I most often hear is how bland and tasteless the food is here in the United States. This comes as no surprise to me. I spend five-plus months each year traveling all over North America, and the thing that bothers me the most is not the hassles at airports, the cramped conditions on airplanes, or the monotony of hotel rooms—it is the food! You see, my family and I grow nearly all the food we consume. It comes from fertile, healthy soils. Your body knows when food is nutrient-dense. It tastes different. It satiates you!

Add to this the fact that synthetic herbicides, pesticides, and fungicides sprayed on most agricultural acreage dramatically reduce nearly every life form in the soil, including fungi, nematodes, protozoa, algae, mites, and microarthopods, as well as earthworms, ants, and other beneficial insects. In 2016, 94 percent of all soybeans grown in the United States were genetically modified, 92 percent of all corn, 95 percent of sugar beets, and 87 percent of the canola.

Quantifying the Impact
of Regenerative Methods

Naysayers often question whether regenerative agriculture can really heal our planet while producing nutrient-dense food. To attempt to answer that question, I agreed to allow LandStream, a consulting company, to quantify ecosystem function on our ranch. LandStream's founders are Abe Collins, a Vermont grazier and consultant, and John Norman, a retired environmental biophysicist who, over a fifty-year career, had invented numerous breakthrough environmental sensors and models for understanding and quantifying landscape function.

With a deep-topsoil future in mind, LandStream supports land managers as they work to heal the land and grow food, fuel, and fiber, to scale regeneration across watersheds and continents, and to couple economics with regenerative management.

LandStream offers support of four functions:

1. Decision support for optimized grazing and cropping management, providing graziers with tracking and forecasting of paddock-by-paddock biomass accumulation, soil moisture, and energy and water fluxes.
2. Quantification of farm and ranch production of ecosystem services, such as reduction of flooding and drought, groundwater recharge, improved stream baseflow, clean water provision, improved soil structure, and improved nutrient cycle along with soil organic matter.
3. A global learning-machine *and* social network that connects farmers and ranchers to facilitate sharing of useful information for achieving improved soil health and landscape function.
4. Engaging the insights and resources of the scientific research community in support of the regenerative

agriculture movement using the universal language of math and environmental biophysics.

LandStream infrastructure ranges from satellite remote sensing to practical ground-level monitoring hardware that tracks solar radiation, weather, soils, vegetation, surface waters, and groundwater. These data are ingested, connected, and analyzed by environmental biophysical models that simulate landscape function. Remote-sensing models used by LandStream quantify per-paddock biomass, land-surface energy fluxes, and the evapotranspiration and soil moisture components of the water-balance.

I decided to work with LandStream because of the need to "land-truth" the data about my farming methods. As producers, we need to show the world that our management makes a difference. In October 2017, we completed the first step in land-truthing at Brown's Ranch by taking one hundred and nineteen soil samples from zero to four feet deep using the Soil Information System of mapping. Some samples showed that we have grown topsoil to a depth of 29 inches, and the soil was well aggregated down to over 36 inches. We can now begin to quantify the outcomes of decades of regenerative management.

The landscape-function quantification system we are implementing at Brown's Ranch will render it a calibration point that thousands of other ranches will be able to make use of. As we begin to stream and model data, we will be quantitatively linking ranch management and outcomes, quantifying ranch production of ecosystem services, and calibrating remote sensing forage tracking and forecasting capabilities. The goal is to help regenerative farmers and ranchers to even more effectively heal and nurture the land in their care.

Cultivation of GMO crops leads to applying higher and higher amounts of these biocides to the landscape, and many of them end up in the air we breathe and the water we drink.

The Environmental Protection Agency's definition of biocides is illustrative: They are "a diverse group of poisonous substances including preservatives, insecticides, disinfectants, and pesticides used for the control of organisms that are harmful to human or animal health or that cause damage to natural or manufactured products." A biocide can't distinguish between an organism that is harmful and one that is beneficial. It kills both equally efficiently. The result is the collapse of life in the soil. Once biology is killed, the soil no longer has the ability to supply the plants with all of the nutrients needed. The plant, then, will not be able to provide animals or humans the nutrients they need.

In *The One-Straw Revolution*, philosopher-farmer Masanobu Fukuoka stated it correctly: "Food and medicine are not two different things; they are the front and back of one body. Chemically grown vegetables may be eaten for food; but they cannot be used for medicine."

In *What's Making Our Children Sick?*, Dr. Michelle Perro and Dr. Vincanne Adams present a sobering thought: "If we are looking for evidence that our food system has failed us, we should look at the children. We have a generation of children whose chronic illnesses do not resemble those of previous generations. Our kids are sicker than their parents, and arguably sicker than their parents were when they were children, regardless of our agricultural and pharmaceutical advances. Clinical evidence indicates that we are doing something wrong. Quite possibly what we are doing wrong today started with the changes to our food production that began just before most of our children were born." Why do the vast majority of producers continue pursuing the Band-aid approach of the current production model? I suspect there are several reasons. The first is fear—fear of letting go of the safety net of familiarity. We are now two generations removed from a production model that did not rely on extensive use of synthetics. Most farmers do not have the experience and knowledge to farm otherwise. A second obstacle to change is the lack of financing options for farmers who move to a different production model. Most lending institutions are not familiar with regenerative agriculture and are hesitant to make loans to very diverse, stacked enterprise operations unless good records or adequate collateral are offered. This is an important reason why I strongly recommend starting small and growing from the profits.

Peer pressure is a huge deterrent to change, too. Most producers might believe they would not be swayed by what others think, but that is simply not true. Those of us who go against the grain have to have thick skin. I often say that I can no longer attend the local livestock auction. If I were to walk in, the place would fall silent because they would have nothing to talk about!

I am not suggesting that America's farmers should immediately eliminate use of any and all inputs. But we have to shift to using inputs judiciously. We have to think through each decision. It is way too easy to say, *I'm going to apply this pesticide today, but I won't use it next time.*

Recent scientific discoveries linking nutrients in healthy soil to regenerative agricultural practices and human health mean we can reverse the downward trend in our children's health. The link is a healthy soil ecosystem—soil biology and the diverse, mutualistic relationships between microbes and plants that help transfer nutrients from the soil into the plants and eventually into us. As this link between healthy soil and our food becomes clear to health-conscious consumers, it presents a tremendous opportunity for farmers and ranchers who switch to regenerative practices. By creating and maintaining biologically active, nutrient-rich soil, those growers will be able to market nutrition instead of commodities, just as we have done with our Nourished by Nature business.

The opportunity to restore nutrient quality and density to everyday food was not even on my radar when I first started changing my farming practices, but I've come to realize that it's one of the important ways I can capitalize on all the work we've done to build healthy soils at Brown's Ranch. As various medical crises in this nation deepen and the costs of health care and treatment rise, more people are looking to their foods for solutions to their illnesses, turning to regenerative and organic farms and voting for biology over chemistry with their pocketbooks. The foods we put in our mouths have the ability to either heal us or harm us. The choices are ours.

Focus on Profit Per Acre

Returning to Paul's comment about outproducing our environment: I had spent so much time chasing yield and pounds, I had not paid enough attention to *profit*. I needed to look at *profit per acre* instead of yield per acre or pounds per calf. Don Campbell's quote echoed in my mind, "If you want to make small changes, change how you do things, if you want to make major changes, change how you SEE things!" From that day on, I focused on profit per acre instead of yield or pounds.

One of the keys to increasing profit per acre is stacking enterprises. The current production model focuses on specialization. Many producers only grow one or two crops. Others have only dairy or only hogs. This has led to them being very efficient at one or two things, but their efficiency comes at a huge cost—the cost of being resilient to either swings in prices or just plain low commodity prices. Not to mention the fact that low diversity has a negative impact on the ecosystem.

Once, when I was speaking to a large crowd of corn and soybean producers in Nebraska, I asked how many of them made a profit on their corn the previous year. One person raised his hand. Yes, only one. I asked how many planned on planting corn the following year. *Every* hand went up.

This is an example of how entrenched people are in today's production model. The only way those producers could make a profit would be to decrease expenses, increase yield, or hope for a major drought in an area where a lot of corn is grown. A drought would lead to a shortage of supply, which would lead to an increase in price. When was the last time input costs decreased? It isn't likely to happen. An increase in yield is possible, but how much would it cost to make that increase happen? A drought is always possible, but unless it was a major drought like that of 2012, it would not affect prices much. As producers, we constantly hear the message that we have to produce more, more, more to feed the world, but we continually suffer with low commodity prices! We need to wake up and realize that there is no shortage of food in the world. There are political and social factors that prevent foods from reaching the hands of people who need it. And there are increasing problems with the lack of nutrient density in the foods produced. But there is *no* shortage of food. Several recently released reports showed that worldwide food production in 2016 was enough to feed 10 billion people. The world's population at the time of this writing is 7.8 billion. If you are waiting on that shortage to increase prices, you will be waiting a long, long time.

Many producers ask me how I increased my profit per acre. The answer is through diversity—diversity of enterprise, which is illustrated in what I call the Brown's Ranch Cash Flow Statement.

As you can see, on our ranch everything revolves around carbon. We grow annual cash crops such as corn, peas, spring wheat, oats, barley, cereal rye, hairy vetch, winter triticale, and a myriad of heirloom vegetables. The grains are run through a quick cleaner, which removes any cracked or broken seeds along with any weed seeds or chaff. The clean grain is sold either as seed or feed to those wanting non-GMO feed (only rarely do I sell grain as a commodity). I am always trying to add value.

The way to make a healthy profit is by taking the waste stream from one enterprise to fuel the profit in another. We feed the screenings from our grain, which we would be docked for if we sold our grains at the elevator, to our laying hens, broilers, and hogs. Thus,

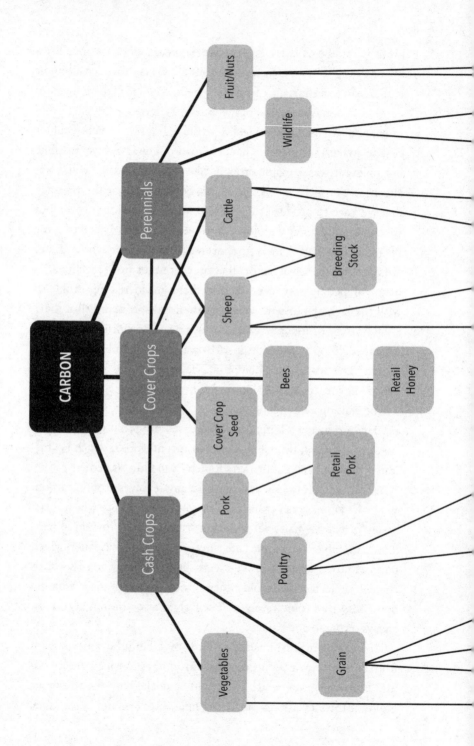

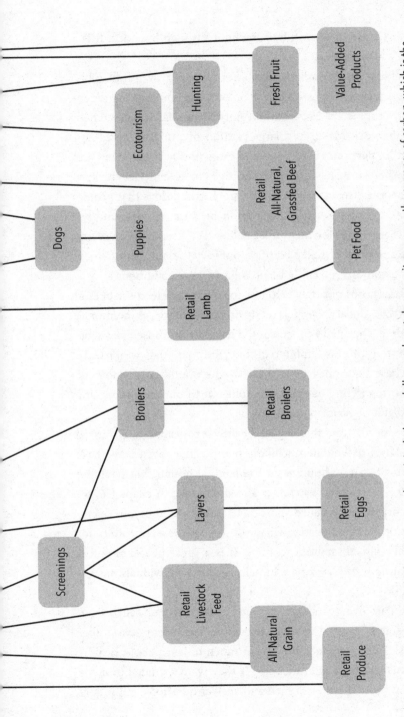

The Brown's Ranch Cash Flow Statement doesn't include any dollars and cents because it measures the currency of carbon, which is the true base unit of value for regenerative farming. With the aid of sunlight, water, and soil biology, plants capture carbon and turn it into our livelihood.

we convert waste into cash by running it through livestock. It is a win-win-win enterprise.

Diversifying our crop base is another way to grow profitability, and thus a relatively new endeavor on our ranch is growing perennial food crops such as fruits and nuts. Perennial forages where we run livestock make up a large portion of our land base. As I described in part one, beef cows and grass-finished beef, ewes and grass-finished lamb, pastured pork, laying hens, and broiler chickens are all income streams. Even the livestock guard dogs that protect the sheep and poultry make us a profit because we raise and sell puppies. Same goes for the border collies. We keep the size of the cowherd and flock of sheep constant but vary the number of stockers and custom-grazed cattle according to forage conditions.

Our cover crops not only feed livestock, but they feed the bees in the hives that a local apiary has set up on our land, too, as mentioned in chapter 6. The apiary processes the honey from those hives separately, putting the raw, unfiltered honey into jars that we provide. We purchase the honey from them by the pound, at a wholesale price. Our marketing business then offers it to our customers, at a healthy profit of course.

But we don't stop there. The healthy ecosystem has spurred an abundance of wildlife, including many game species. We have chosen not to allow hunters on the land, although that could be a nice income stream. Instead, as also mentioned in chapter 6, we allow an organization called Sporting Chance to hunt on our operation free of charge. Sporting Chance provides the opportunity for individuals with disabilities to hunt. It is a great program and is very rewarding to play a part in helping those individuals harvest an animal.

In the past five years, we have had visitors to our ranch from all fifty states, every Canadian province, and twenty-two other countries. These people come to our ranch to learn about healthy ecosystems and how soil functions. If they want a guided tour, we will gladly show them our ranch and answer questions, but it does take a lot of time. (We had over 2,500 visitors in 2017.) Because it is a strain on our time and we have work to do, we charge them a fee. This pays us for our time. Another income stream.

Because Shelly and Paul and I have set up this rich diversity of income streams, our ranch is resilient not only ecologically but financially, too. I will take profit over yield any day. It is much more enjoyable to sign the back of the check and not the front!

When I travel around North America to give presentations about our farming methods, I hear over and over again that there is no money in production agriculture. My story, however, is proof that there is good money to be made when you think outside the box.

Even though we already run seventeen enterprises on our operation, we have many more that we would like to add in the future. Among the possible additions are rabbits, turkeys, goats (oops, actually we just added some meat goats to the operation), a food truck, cheese making, and soap making.

At almost every presentation I give, someone asks: "How many employees do you have in order to get all of that work done?" The answer is: Shelly, me, Paul, and Paul's girlfriend, Shalini Karra. For five or six months each year, we bring on a couple of interns. As I described in *Teaching the Younger Generation* on page 90, working with interns is a lot of fun, and they certainly do help with the farm work, but their training also requires an investment of time, because most of them have no farm experience to start with.

Rather than tallying up how many enterprises we run and thinking of it as a burdensome work load, I tell people to think about all of the tasks we don't need to do on Brown's Ranch because we let nature do them for us. For example, we don't have to haul and apply fertilizer, pesticides, and fungicides. We don't need to vaccinate and worm our livestock. We don't spend days chasing around the country to find the latest and greatest bulls, rams, and boars. We don't pregnancy test the cowherd, pigs, or sheep. We don't have daily chores of starting up farm equipment to haul feed to the livestock during the winter. We don't have to spend time hauling manure from the corrals out to spread on the fields—come to think of it, we don't even have to spend the time repairing corrals! I could go on, but I'm sure you have gotten my point.

Another common question is: How many acres does it take to be profitable? I reply that it is not the number of acres that matters. I have seen many profitable operations that operate on less than an acre of land. Stacking enterprises gives greater opportunity for profitability. Anyone can be profitable on their land base if they are willing to avoid the pitfalls of the current production model, focus on regenerating their ecosystem, and strive for profit not yield.

Conclusion

Do Something

I wrote this book to tell the story of how my family and I changed from farming and ranching with an "industrial" mindset to farming and ranching in nature's image. This book is not intended to tell you how you should, or should not, run your farm, ranch, or garden. Only you can decide that.

For me, learning to farm and ranch well has been a thirty-plus-year journey. One in which I first had to unlearn so that I could relearn. I hope that the fundamental points I put forward in this book have come through clearly and convincingly. The most important one is that all of us—whether farmer, rancher, or home gardener—have the ability to harness the awesome power of nature to produce nutrient-dense food. We can do this in a way that will both regenerate our resources and ensure that our children and grandchildren have the opportunity to enjoy good health. For me, this journey has been one of following the agricultural principles I have outlined in this book, along with my own personal principles.

Trust God. When Shelly and I went through those four years of disasters, we were about as broke as one could be. I tell people that we were so broke that the banker knew when we bought toilet paper. Now that is broke! But we had each other, and we had faith. Proverbs 3:5–6 says, "Trust in the Lord with all your heart and lean not on your own understanding; in all your ways submit to him, and he will set your path straight."

We did not know what God had planned for us, but we knew he had a plan. Faith is critical as you move down the regenerative path. You have to trust that soil biology will improve if you apply the five principles. You have to trust that the nutrient cycle, the energy cycle, and the water cycle will improve. You can rest assured because God would not create an imperfect system. The proof is in the eons of time that natural ecosystems have functioned regeneratively.

All I ever wanted to do is ranch. It was never my desire to spend the majority of my life traveling the world sharing my story. Yet that is what I do. I truly believe that God put Shelly and me through those four years of disaster in order to use us to help, in some small way, to heal this planet. Think of the odds. What are the odds of losing four crops in a row to hail and drought when none of my neighbors suffered four years of losses?

Keep an open mind. Be willing and open to learning. I can't begin to tell you the number of people who have said, "But Gabe, you don't understand. We can't do that here. It just won't work here!" That is their perspective; my perspective is that they were simply not ready to learn! Henry Ford said it best: "If you think you can or you think you can't, you are correct." The majority of people who attend one of my presentations have their mind made up before I even begin to speak. One of the things that helped me is I did not grow up on a farm, I grew up in town. When I entered production agriculture I did not have any preconceived ideas. I had an open mind and was ready to learn. Unfortunately, first I learned the conventional production model, so then I had to unlearn and learn again. Remember what Don Campbell taught me, "If you want to make small changes, change how you do things; if you want to make major changes, change how you see things."

Observe. Job 12:7–8 says, "But ask the beasts and they will teach you, the birds of the heavens and they will tell you, or speak to the earth and it will teach you, or let the fish in the sea inform you."

During those four years of disasters, I found great solace in spending time walking the pastures and fields. I learned to observe.

The sweet scent of clover on a summer breeze. The gentle rustling of the cattle as they graze. Notice how their whiskers touch a plant before they bite it. Observe the grasshoppers—why are they feeding on the thistle? Because it is lower in nutrients. Grab a handful of soil, is it well aggregated? Why does it smell pungent? That is the actinomycetes, which means the soil is bacterially dominant. Take a shovel out into your field or garden; does it sink into the soil easily? Notice how the oat roots are traveling horizontally. That's a sign of a restrictive layer in the soil due to a tillage pass years ago. See all the dandelions? They are a sign that the soil lacks calcium and is high in potassium.

I think that using our senses is a forgotten art. The obsession with immediate profit erases our connections with our ecosystem, and without those connections, our presence on the land becomes exceedingly rare. In the eight years I worked alongside my father-in-law, I never once saw him put a shovel blade in the soil and turn up the earth to take a good look at it. In four years of agricultural studies in college, not once did any of my professors ever tell me—let alone teach me—how to observe. And without that skill, we cannot farm, ranch, or garden in nature's image.

Do not be afraid to fail. Henry Ford said, "Failure is simply the opportunity to begin again, this time more intelligently." I could write several books about all my failures. I often say that I had to fail at everything twice, usually the hard way. But I learned from those failures, and that is what is important. Remember the story about how I seeded perennials directly into severely degraded soil without first priming the soil by planting cover crops? I have not made that mistake again. I now always seed cover crops into a field for at least two years before seeding it to perennials.

In October 2016, we placed most of our hay out in our perennial pastures, leaving very little near the farmstead. Our plan was to have the animals bale-graze it during February and March. Winter hit early, and by the first week of January, over one hundred inches of snow had fallen. In those conditions, it cost us a lot of time and money to move the cattle out to the hay and to move some hay back to the farmstead. We won't make that mistake again! I tell people who visit our ranch that we try hard to fail at several things every year. For example: How will I know whether a certain cover crop species will work in my environment unless I try planting it? It might turn out to be a total loss, but how else can I learn? Our ranch is much better off today because of our failures.

Understand your context. In *The Unsettling of America*, Wendell Berry wrote: "While we live our bodies are moving particles of the earth, joined inextricably both to the soil and to the bodies of other living creatures. It is hardly surprising, then, that there should be some profound resemblances between the treatment of our bodies and our treatment of the earth."

We must understand our social, ecological, and spiritual context.

Simply put: We have to understand that every single thing we do has compounding and cascading effects. If we use tillage, synthetic fertilizers, pesticides, fungicides, wormers, vaccines, or any other disturbance, we will be impacting *all* ecosystems—the soil, the water, the air, and society. We do nothing in singularity. As gardeners, farmers, and ranchers, we are producing nutrition, the very nutrition that defines human health. We have to take responsibility for our actions. We cannot "turn a blind eye." What effect do our actions have on our families? Our communities? Our ecosystems? Our relationship with God?

Do something. Last spring, I was out in the field seeding a diverse polyculture cash crop when my phone rang. I answered, "Hello, this is Gabe." No one replied, but I could hear talking in the background. I repeated, "Hello, this is Gabe." "Oh, my goodness! It is him, he answered!" a voice on the other end exclaimed. "Yes, it is," I replied. "I can't believe you answered the phone," she retorted. "Why wouldn't I?" I responded. "Because you're Gabe Brown!" she excitedly said. I laughed and told her that talking to me really wasn't a big deal. She gave me her name and asked if I had time to answer a few questions about gardening. I told her that I would be seeding for about another half hour before I would have to stop and fill the drill, so I was all ears. She explained to me that she was from inner city Detroit and she desperately needed to grow food for the area children, many of whom were suffering from malnutrition. My heart sank as she continued to explain that the only meal many of these children received each day was the noon lunch provided at school, if they were old enough to attend. She told me how she was spending what little extra money she had on buying food for the youngest of the children. She did not have the money to buy food for those in school, but she had the idea of growing some vegetables. If she could just grow some vegetables, she would have some food for them.

I stopped the tractor and shut it off. She explained that there were several vacant lots next to her house that were overgrown with weeds. She wanted to know if I thought it was possible to grow vegetables on them. For the next hour and a half, I explained how to make compost from waste, how to mix that compost with any soil she could find, where to source seed, which varieties, how to use cardboard and newspapers to inhibit weeds, and any other information I thought she might find useful. I then told her to call me if she ever had any questions, wished her well, and thanked her for *doing something!*

We ended the call, and as I put my phone away, I realized why God put me on the journey of dirt to soil: *God created you, so do something!*

Acknowledgments

Support for portions of this book was provided by the Leopold Center for Sustainable Agriculture at Iowa State University, Mark Rasmussen, director. Additional support was provided by: the Regenerative Agriculture Foundation, the Lydia B. Stokes Foundation, the Community Foundation for San Benito County, Dennis and Trudy O'Toole, and the New Society Fund. Thanks to everyone!

Recommended Resources

Books

Berry, Wendell. *The Unsettling of America: Culture and Agriculture.* Berkeley, CA: Counterpoint, 1996.

Brunetti, Jerry. *The Farm as Ecosystem: Tapping Nature's Reservoir— Biology, Geology, Diversity.* 2nd ed. Austin, TX: Acres USA, 2014.

Clark, Andy. *Managing Cover Crops Profitably.* 3rd ed. College Park, MD: SARE Outreach, 2007.

Davis, Walt. *How to Not Go Broke Ranching: Things I Learned the Hard Way in Fifty Years of Ranching.* North Charleston, SC: CreateSpace Independent Publishing Platform, 2011.

Faulkner, Edward. *Plowman's Folly.* Reprint ed. Norman, OK: University of Oklahoma Press, 2012.

Fukuoka, Masanobu. *The One-Straw Revolution: An Introduction to Natural Farming.* New York Review Books Classics, 2009.

Hines, Chip. *How Did We Get It So Wrong: The Reality of Ignoring Nature.* North Charleston, SC: CreateSpace Independent Publishing Platform, 2012.

Howard, Albert. *An Agricultural Testament.* Oxford, UK: Benediction Classics, 2010.

Lewis, Meriwether, and William Clark. *The Journals of Lewis and Clarke* (selections), edited by Bernard DeVoto. Boston: Houghton Mifflin, 1953.

Miller, Daphne. *Farmacology: Total Health from the Ground Up.* New York: William Morrow, 2013.

Montgomery, David. *Growing a Revolution: Bringing Our Soil Back to Life.* New York: W. W. Norton and Company, 2017.

Niman, Nicolette. *Defending Beef: The Case for Sustainable Meat Production.* White River Junction, VT: Chelsea Green, 2014.

Perro, Michelle, and Vincanne Adams. *What's Making Our Children Sick?: How Industrial Food Is Causing an Epidemic of Chronic Illness, and What Parents (and Doctors) Can Do About It* White River Junction, VT: Chelsea Green, 2017.

Salatin, Joel. *Fields of Farmers: Interning, Mentoring, Partnering, Germinating.* Swoope, VA: Polyface Farms, 2013.

Savory, Allan, with Jody Butterfield. *Holistic Management: A Commonsense Revolution to Restore Our Environment.* 3rd ed. Washington, DC: Island Press, 2016.

Schatzker, Mark. *The Dorito Effect: The Surprising New Truth About Food and Flavor.* New York: Simon & Schuster, 2015.

Schwartz, Judith. *Cows Save the Planet: And Other Improbable Ways of Restoring Soil to Heal the Earth.* White River Junction, VT: Chelsea Green, 2013.

Sinek, Simon. *Start with Why: How Great Leaders Inspire Everyone to Take Action.* New York: Penguin, 2009.

Wilson, Gilbert. *Buffalo Bird Woman's Garden: Agriculture of the Hidatsa Indians.* St. Paul: Minnesota Historical Society Press, 1987.

Websites

Amazing Carbon: http://www.amazingcarbon.com

Bionutrient Food Association: http://www.bionutrient.org

EcoFarmingDaily e-newsletter: http://www.ecofarmingdaily.com

Holistic Management International: https://holisticmanagement.org

Savory Institute: http://www.savory.global

Soil Foodweb Inc.: http://www.soilfoodweb.com

Soil Health Consultants, LLC: http://www.soilhealthconsulting.com

Solutions Beneath Our Feet: Can Regenerative Agriculture and Healthy Soils Help Combat Climate Change?: http://www.youtube.com/watch?v=XlB4QSEMzdg

Winona Ranch: http://www.pasturecropping.com

Index

Note: Tables are indicated by an italic *t* following the page number.

About the Author

Gabe Brown is a pioneer of the soil-health movement and has been named one of the twenty-five most influential agricultural leaders in the United States. Brown, his wife, Shelly, and son, Paul, own Brown's Ranch, a holistic, diversified 5,000-acre farm and ranch near Bismarck, North Dakota. The Browns integrate their grazing and no-till cropping systems, which include cash crops and multispecies cover crops along with all-natural, grass-finished beef and lamb, pastured pork, and laying hens. The Brown family has received a Growing Green Award from the Natural Resources Defense Council, an Environmental Stewardship Award from the National Cattlemen's Beef Association, and the USA Zero-Till Farmer of the Year Award.